Media Events in WEB 2.0 CHINA

THE SUSSEX LIBRARY OF

Asian and Asian American Studies

Series Editor: Dr Mina Roces, School of History, The University of New South Wales

The Sussex Library of Asian Studies publishes academic manuscripts in various disciplines (including interdisciplinary and transnational approaches) under the rubric of Asian Studies – focusing on Economics, Education, Religion, History, Politics, Gender, and comparative studies with the West and regional studies in Asia.

China's Rising Profile: The Great Power Tradition, Harsh V. Pant, King's College London.

Chinese Identity in Post-Suharto Indonesia: Culture, Media, Religion and Language, Chang-Yau Hoon, Singapore University.

Dancing the Feminine: Gender & Identity Performances by Indonesian Migrant Women, Monika Swasti Winarnita, University of Victoria, BC, Canada.

Family Ambiguity and Domestic Violence in Asia: Concept, Law and Process, edited by Maznah Mohamad, National University of Singapore, and Saskia E. Wieringa, University of Amsterdam.

Han Shan, Chan Buddhism and Gary Snyder's Ecopoetic Way, Joan Qionglin Tan, Hunan University, China and University of Wales, Lampeter.

Heteronormativity, Passionate Aesthetics and Symbolic Subversion in Asia, Saskia E. Wieringa, University of Amsterdam, with Abha Bhaiya and Nursyahbani Katjasungkana.

The Independence of East Timor: Multi-Dimensional Perspectives – Occupation, Resistance, and International Political Activism, Clinton Fernandes, University of New South Wales.

Negotiating Malay Identities in Singapore: The Role of Modern Islam, Rizwana Abdul Azeez, Institute of Southeast Asian Studies.

Media Events in Web 2.0 China: Interventions of Online Activism, Jian Xu, University of New South Wales, Australia.

The Politics of Dress in Asia and the Americas, edited by Mina Roces and Louise Edwards, University of New South Wales, Sydney and University of Technology, Sydney.

Pool of Life: The Autobiography of a Punjabi Agony Aunt, Kailash Puri (co-author of *The Myth of UK Integration*), and Eleanor Nesbitt, University of Warwick.

Southeast Asian Migration: People on the Move in Search of Work, Marriage and Refuge, edited by Khatharya Um, University of California, Berkeley, and Sofia Gaspar, CIES-IUL, Lisbon, Portugal.

Media Events in WEB 2.0 CHINA

INTERVENTIONS OF ONLINE ACTIVISM

Jian Xu

sussex
ACADEMIC
PRESS
Brighton • Chicago • Toronto

2 4 6 8 10 9 7 5 3 1

First published 2016 in Great Britain by
SUSSEX ACADEMIC PRESS
PO Box 139
Eastbourne BN24 9BP

Distributed in North America by
SUSSEX ACADEMIC PRESS
Independent Publishers Group
814 N Franklin St, Chicago, IL 60610, USA

British Library Cataloguing in Publication Data
A CIP catalogue record for this book is available from the British Library.

Library of Congress Cataloging-in-Publication Data
Names: Xu, Jian (Teacher of Chinese), author.
Title: Media events in web 2.0 China : interventions of online activism / Jian Xu.
Description: Brighton ; Chicago : Sussex Academic Press, 2016. | Series: The Sussex library of Asian and Asian American studies | Includes bibliographical references and index.
Identifiers: LCCN 2015034924 | ISBN 9781845196356 (hb : alk. paper)
Subjects: LCSH: Internet—Political aspects—China. | Internet—Social aspects—China. | Mass media—Political aspects—China. | Mass media—Social aspects—China. | Online journalism—China. | Citizen participation—China.
Classification: LCC HN740.Z9 I56895 2016 | DDC 302.23/10951—dc23
LC record available at http://lccn.loc.gov/2015034924

Typeset and designed by Sussex Academic Press, Brighton & Eastbourne.
Printed and bound by CPI Group (UK) Ltd, Croydon, CR0 4YY

Contents

Series Editor's Preface

The Sussex Library of Asian and Asian American Studies Series publishes original scholarly work in various disciplines (including interdisciplinary and transnational approaches) under the rubric of Asian studies—particularly Economics, Education, Religion, History, Politics, Gender, Comparative Studies with the West, and Regional Studies in Asia. The Series is keen to publish in emerging topics that demand attention in the Asian context—from the politics of dress to the heteronormative in India and Indonesia for example. Seminal works and approaches will find a home here. The Series also welcomes single-country studies or anthologies that explore one important theme across a number of Asian contexts. We expect the Series to contribute to the scholarly debates on topical issues, highlighting the importance of the region. From 2015 onwards we are expanding our repertoire to include new perspectives in the vibrant field of Asian American studies.

In *Media Events in WEB 2.0 China Interventions of Online Activism*, Jian Xu presents a fascinating and insightful study of the development of media events since 1949 with an emphasis on China's internet era. This pioneering work shows how the practices of online activism of ordinary Chinese have been able to transform China's media events from the bottom up. This book shifts the scholarly perspective from examining China's internet activism in transnational spaces to an analysis of how online activists at home have been able "to produce nonofficial discourses, generate networked public opinions and take collective action", resulting in transforming media events into spectacles that are "contentious, multi-voiced, participatory and deliberative". Xu's main arguments come from the analysis of case studies of three media events: the CCTV Spring Festival Gala, citizen journalism in the 2008 Sichuan earthquake, and the online *weiguan* or virtual crowd gathering as networked collective action in scandalous media events. His primary sources that include interviews, participant observation and textual analysis are mined to produce the first work to use a 'media event' framework to examine China's online activism and internet politics in the boundaries of the nation state. Given that China's online population makes it the world's largest internet market, and that "the Internet has become the most important alternative media in present-day China," Xu's interpretations will be crucial to understanding much of activist policies in China today. At the same time the 'media events' approach suggests an excellent model that could be applied to societies around the world.

List of Tables, Figures and Illustrations

Acknowledgments

This book started as a doctoral research project in the School of Arts & Media at the University of New South Wales, and was completed during my Endeavour Postdoctoral Research in the Faculty of Arts and Social Sciences at the University of Technology, Sydney. I would like to express my utmost gratitude to Professor Gerard Goggin, Dr. Haiqing Yu and Associate Professor Elaine Jeffreys for their guidance and support during my doctoral and postdoctoral research. Their intellectual insights and rigorous scholarship have not only inspired this book, but also my plans for an academic career. My hearty appreciation also goes to Professor Mina Roces, Series Editor of the Sussex Library of Asian and Asian American Studies, and Anthony Grahame, Editorial Director at Sussex Academic Press, for their generous assistance and high efficiency in the production of the book.

I would like to express my deepest thanks to Dr. Mu Yang, Professor Wanning Sun, Professor Michael Keane, Professor Yanan Zheng and Ms. Xing Jin for the opportunities, suggestions and support they have given me in my research and academic career. I am grateful to Associate Professor Guobin Yang for the enormous intellectual contribution of his research and his generous support during my visiting research in the Annenberg School for Communication and the Centre for the Study of Contemporary China at the University of Pennsylvania.

I remain forever grateful to my family: Yuqing Wang, Guoquan Xu and Guanxiong Wang for their unconditional love and constant support for all of my endeavours in life. I dedicate this book to them.

Earlier versions of some parts of Chapter 2 and Chapter 4 of this book have been published in Kevin Howley (ed.), *Media Interventions* (New York: Peter Lang, 2013): 341–356, and Guobin Yang (ed.), *China's Contest Internet* (Copenhagen, Demark: NIAS Press of the University of Copenhagen, 2015): 257–282. I acknowledge the two publishers for granting permission to reproduce part of the published materials in this book.

Introduction

Since Deng Xiaoping's reform and opening-up policy in 1978, China has witnessed great structural transformations in the economic, cultural, technological and political spheres. The Chinese media, which interacts with these spheres, has played a central role in representing and facilitating these transformations while itself being changed in the process. Since the mid-1990s, in particular, the Internet in China has developed rapidly and profoundly impacted Chinese society. According to the China Internet Network Information Centre (CNNIC), China's Internet population reached 668 million in June 2015 (CNNIC 2015): the largest number of users of any country in the world. The rapid development of the Internet and other digital technologies has created new digital platforms and new communicative forms for Chinese people to express, connect and interact, thereby giving rise to what Manuel Castells (2000) calls the "interactive network" and a "networked society" in China. Thus, the digital revolution has not only created new media cultures, but also transfigured Chinese society.

This book investigates one of the most important forms of China's new media cultures—online activism, as well as its social, cultural and political impacts. Based on my analysis of three types of media events (ritual celebration, natural disaster and political scandal) in China's digital age, I argue that the alternative and activist new media practices of ordinary Chinese people have intervened in and transformed these types of media events, which have conventionally been dominated by the Chinese Communist Party (CCP), mainstream media and the market. By hijacking, appropriating and reinventing media events, non-state actors [1] aided by the Internet are able to produce nonofficial discourses, generate networked public opinions and take collective action, and hence generate bottom-up momentum to make media events contentious, multi-voiced, participatory and deliberative.

From enchantment to disenchantment: The evolution of media events

Dayan and Katz (1992), in their seminal work, *Media Events: The Live Broadcasting of History*, define media events as spectacularized and live-broadcast events, which interrupt the normal rhythm of government, mass media and national audiences by attracting much of their attention. They are "ritual celebrations", which create "periodic social gatherings for the celebration of society" (Rothenbuhler 1998: 79). Dayan and Katz further identify media events as a distinctive genre of media communication on three levels: syntactic, semantic and pragmatic. On the syntactic level, media events are interruptions of broadcasting routines, transmitted live, preplanned and organized outside the media by those within the establishment. On the semantic level, media events are presented with ceremonial reverence and serve to celebrate reconciliation rather than conflict. On the pragmatic level, media events speak to massive national audiences and are necessarily monopolistic and hegemonic (1992: 9–14). Dayan and Katz also distinguish three scripts of media events: contest, conquest and coronation, evidenced by the Olympics, the first human landing on the moon and the wedding of Charles and Diana. All these carefully scripted and live televised events create "liminal moments", through which "social integration of the highest order is thus achieved via mass communication" (1992: 15).

However, the globalization of media industries, proliferation of delivery platforms and fragmentation of television audiences have increasingly challenged the premise that "television plays a central role in governance of the nation-state", which Dayan and Katz's classic definition of media events operates on (Sun 2014: 458). Therefore, scholars propose using a decentralized ritual view (Couldry 2003), extended typification (Katz and Liebes 2007) or a broadened definition (Cottle 2006; Scannell 1999) to revise the classic concept of media events. Hepp and Couldry, in their introduction to the edited volume *Media Events in A Global Age*, systematically review the existing critiques of media events by considering three main aspects: the ritual perspective, narrowness of three typified scenarios and the core definition of the genre (2010: 5–8).

First, Dayan and Katz take media events as rituals of affirmed values and fail to account for their conflicting and diverse characteristics in post-modern and globalized societies (Hepp 2004). In order to overcome this problem, Nick Couldry suggests a focus on "the ritual space of media", which is "the opposite of isolating particular moments and elevating them to special, even 'magical' significance", and instead involves "tracing the antecedents of media rituals in the patterns, categories and boundaries at work everywhere" (2003: 13). He suggests shifting the focus from the integrative role of ritual media events to the "media rituals" that construct the "myth of the mediated centre" through various forms of media communication (2003: 13). This de-centered ritual perspective views media events as being constituted by different forms of communication, which "articulate the power-related, hegemonic imagination of the media as the centre of present societies, as the expression of the important incidents within that society" (Hepp and Couldry 2010: 5).

Second, Dayan and Katz's three basic types of media events are too narrow to cover the increasingly event-based nature of media representations. Scannell (1996) argues that Dayan and Katz omit "major news events" that are unexpected, disruptive, and shock the world. Katz and Liebes (2007) point out that major changes in technology, the organization of broadcasting institutions and the credibility of governments have led to a reduction in both the frequency and centrality of ceremonial media events. They suggest extending the typification of ritual media events to include conflict-oriented media events, such as disasters, wars and terrorist attacks (Katz and Liebes 2007). The major difference between ceremonial and disruptive media events is that the latter is characterized by "an overtaking of the public domain by oppositional forces (not hegemonic ones)" (Hepp and Couldry 2010: 7).

Third, the core definition of media events as genres has been challenged. Scannell (1999; 2002) distinguishes natural "happenings" with human-made "events", and proposes to take media events as mediatized performances in specific historical contexts. Douglas Kellner (2003) proposes the concept of "media spectacle", a broader term than "media events", to describe social, cultural and political dramas and conflicts that are represented in mass media. According to Kellner, media spectacles are all "spectacular media events" which interrogate contemporary politics, culture, society and everyday life (2003: 1). Simon Cottle (2006) suggests extending the classic notion of media events to look at "mediatized rituals". Mediatized rituals refer to "those exceptional and performative media phenomena that serve to sustain and/or mobilize collective sentiments and solidarities on the basis of symbolization and a subjunctive orientation to what should or ought to be" (Cottle 2006: 415). By using the instructive approach, Cottle (2006) accommodates different genres of media events under the overarching concept of mediatized rituals and further categorizes media events into six sub-classes: celebratory media events, moral panics, conflicted media events, media disasters, media scandals and mediatized public crises.

Daniel Dayan also reflects upon and revises his original approach towards ritual media events in his conclusion to the edited volume *Owning the Olympics: Narratives of the New China* (Price and Dayan 2008). He argues that the Beijing Olympics are no longer monopolized by national broadcast media as the Olympics were in his original work, but are staged and scrutinized in a newly-volatile global media environment. The dramatic expansion of media outlets and the growth of mobile communication technologies have made it significantly more difficult to regulate or control the meanings of media events. As Dayan puts it:

> This kind of exclusive focusing on one event at any given time is now becoming almost impossible. Instead, there is a 'field' of events in which different candidates compete with each other for privileged status, with the help of entrepreneurial journalists. Social and political polarization and its effect on media mean that it is harder to achieve a broad consensus about the importance of particular events. News and media events are no longer starkly differentiated entities but exist rather on a continuum. This banalization of the format leads to the emergence of an intermediate zone characterized by the proliferation of what I would call 'almost' media events. (2008: 396)

Dayan (2008) further identifies the characteristics of media events in the post-broadcasting era as "disenchantment, derailment and disruption". Media events do not necessarily enhance social integration or consolidate shared values as in the broadcasting era. Instead, they may be "subverted (denounced)", "diverted (derailed)" or "perverted (hijacked)" in globalized media cultures (Dayan 2008: 399).

Inspired by these critiques of media events as well as Dayan's update on his original work, Hepp and Couldry suggest a decentred and populist approach to studying ritualized media events in the global age, and redefine media events as "certain situated, thickened, centering performances of mediated communication that are focused on a specific thematic core, cross different media products and reach a wide and diverse multiplicity of audiences and participants" (2010: 12). First, they believe media events should be understood as "thickenings of media communication", because in the post-broadcasting era they are co-produced by media organizations across nations through different media channels, media products and communicative forms. Second, they argue media events are "power-related articulations" and "cannot be related to just one power centre". Media events have opened up space for multiple social actors to establish "discursive positions". Third, they argue that the representations and meanings of media events vary "across different media cultures in its thematic core" (Hepp and Couldry 2010: 11–13).

Hepp and Couldry's redefinition and reconceptualization provides a revised frame through which to understand media events in globalized media cultures. The evolving paradigm of media events, from Dayan and Katz's original work to Hepp and Couldry's revision, illuminates the transformation of contemporary Chinese media events. Since the founding of the People's Republic of China (PRC) in 1949, media events have been evolving alongside the nation's society, culture, politics and technology. From authoritarian media events manipulated by the CCP with ritual functions and affirmed values, they have become more negotiated during the Internet age, which involves multiple media channels, social actors and discursive positions.

This book examines the development of Chinese media events since 1949 with a particular focus on media events in China's Internet era. Rather than looking at the transnational and transcultural expansion of media discourses that Chinese media events embody, I focus on factors internal to Chinese society which have led the remodeling of media events in China. The domestic focus allows me to analyze the complicated interplays between the state, media, market and society in this process. I argue that the practices of online activism of ordinary Chinese people have become the most important internal force that has intervened in and transformed China's media events from the bottom-up. Considering the development of media and communications technologies and China's media systems over the past sixty years enables an appreciation of the extent, nature and progression of this evolution.

Transformation of Chinese media events

Chinese media events post-1949 have been evolving with China's media system and communication technologies. They have provided rich social texts for the investigation of China's society, culture and politics as well as the interrelations among the state, media and society in different historical periods. In order to historicize and contextualize the transformation of Chinese media events, I periodize China's media events into three historical stages according to the characteristics of China's media and communication system in different historical periods. These stages include "authoritarian" media events in the pre-reform era, "marketized authoritarian" media events in reform era, and "deliberative" media events in the new media era. This periodization underscores the relationship between the structural transformation of Chinese media and the functional transformation of Chinese media events.

The pre-reform era and "authoritarian" media events

Prior to the late 1970s, the Chinese media strictly followed a "commandist system associated with Communist ideology" (Pan 2000: 73). They were apparatuses of the CCP's propaganda machine, serving its purposes of indoctrination, mass persuasion, propaganda and social mobilization (Schurmann 1968; Yu 1964). The journalistic practices of the Chinese media were guided by the "Party principle" [*dangxing yuanze* 党性原则]. According to Yuezhi Zhao, the Party principle emphasized three basic requirements: the news media must accept the Party's guiding ideology as its own; they must propagate the Party's programs, policies and directives; and they must accept the Party's leadership and adhere to the Party's organizational principles and press policies (1998: 19). In order to maintain the legitimacy of the Party principle, the Chinese media were tightly controlled and strictly censored by Party propaganda departments on the state, provincial and local levels.

The highly centralized structure of media and communication systems as well as the function of the mass media as the Party's "propaganda machine" and "mouthpiece" in this pre-reform era created "authoritarian media events". They were broadcast via radio and television and propagated via the print media and other popular media, such as political propaganda posters. Similar to those in the Eastern bloc during the Cold War era, media events in pre-reform China were "retrospective commemorations", which followed a "highly structural ideological discourse". They offered displays of "control" or "manifestations of obedience". The performer and respondent were ""ideologically and scenographically interwoven" into a "choreographic oneness" and spoke "in the same voice" (Dayan and Katz 1988: 163–64).

Staging authoritarian media events was one of the most important political tasks of the mass media in pre-reform-era China. In these events, the organizers, media producers and audiences were not "three contractual partners" as in democratic media events in the West (Dayan and Katz 1992). Instead, media producers and audiences cooperated and participated in media events under the manipulation of the organizers (usually the CCP). By organizing authoritarian media events, the Party displayed its absolute power over the media and the people. The legitimacy of the CCP and the social integration of the Communist regime were consolidated through these periodic ritual celebrations. The National Day military parades from 1950 to 1959 and Mao's reception of the Red Guards in Tiananmen Square in 1966 were representative examples of such authoritarian media events. Though alternative media channels existed in the pre-reform era, such as underground radio stations and tabloids, they had very little social impact and did not seriously challenge the gigantic Party media organization because of their illegal status, limited audiences, and the government's strict regulation and censorship.

With the collapse of socialism in Easter Europe in the late 1980s, authoritarian media events decreased in number, frequency, scale and influence. However, they still exist today in totalitarian regimes, as in the funeral of Kim Jong-il in 2011. Though China is still under the CCP regime and maintains socialist ideology, the deepening of market-oriented economic reforms post-1978 has given rise to a series of media reforms, and in the meantime, has transformed authoritarian media events with market forces.

The media reform era and "marketized authoritarian" media events

Since 1978, China has implemented fundamental reforms in its economic system under the leadership of Deng Xiaoping. The planned economy has gradually been replaced by the market economy. The government has also initiated a series of reforms in media and communications, such as media marketization and conglomeration. These reforms have caused not only structural change to China's media landscape but also the transformation of China's media events.

Marketization has allowed media organizations to transition from being non-profit public institutions into economic entities involved in market competition. Increasingly fierce market competition has forced the media to better serve audiences' needs, and thereby enhanced its role as an information and entertainment provider, and sometimes government watchdog (He 1998). Media marketization has weakened the propaganda function of the Chinese media, enlivening the industry and resulting in the multiplicity and diversity of China's media culture (Harrison 2002; Yu 2009). Ostensibly in accordance with marketization principles, the government started to encourage cross-regional and cross-media conglomeration across the state media since the mid-1990s, in order to make China's media industry "bigger and stronger" [*zuoda zuoqiang* 做大做强]. However, in contrast to media conglomeration in the West, China's media conglomeration has seldom resulted from free market competition, but rather from administrative orders, causing Chinese media to be popularly referred to as "Party Publicity Inc" (He 1998).

China's media reforms have resulted in the "dual-track" nature of Chinese media, which is characterized by the "interlocking of Party control and market forces" (Y. Zhao 1998: 3). On the one hand, the market mechanism is harnessed to stimulate the development of the media industry; on the other hand, the Party still controls the assignment of media organizations' top personnel and guides the direction of important media content, such as news and current affairs. Hence, the Chinese media are caught between what Yuezhi Zhao (1998) calls the "Party line" (ideological disciplines) and the "bottom line" (market principles).

Along with these media reforms, China's media events have been remade from "authoritarian" to "marketized authoritarian". In contrast to the "authoritarian" media events in pre-reform China, which were completely dominated by the CCP and the state media, "marketized authoritarian" media events are motivated by market profits and have diverse producers and types. For example, China Central Television (CCTV) has started to produce media events which continue ideological propagation but are also economically profitable, such as the annual Spring Festival Gala [*Chunjie lianhuan wanhui* 春节联欢晚会] which began in 1983. The market mechanism has loosened the Party's control of media content and created a more open and transparent communications environment. The types of media events have extended from being merely celebratory and commemorative in the pre-reform era to include more disruptive media events, in order to satisfy the public's information needs, establish different media's brand images and obtain better economic benefits. Natural disasters, public crises and political scandals are allowed to be reported by central and local media, though still under control, and become national media events, such as the 2003 SARS epidemic, and the 2008 winter storm and Sichuan earthquake.

In the media reform era, media events have been "marketized" and "diversified" by market forces. However, they remain "authoritarian" in nature, because the Chinese media are still constrained by the CCP's propaganda imperative and administrative orders. The "Party line" is always superior to the "bottom line", because "only by serving the party-state's political interest [are they] granted economic privileges" (Lee, He, and Huang 2006: 586).

The changes in China's media events have paralleled not only China's deepening media reforms, but also the development of Information and Communication Technologies (ICTs). The flourishing of ICTs has further decentralized and socialized the Chinese media, and created alternative spaces for ordinary Chinese citizens to participate in media events, thereby transforming media events from being "marketized authoritarian" in nature to "deliberative".

The new media era and "deliberative" media events

In an attempt to gain technological legitimacy, promote economic growth and round out national strength, the government has taken a series of measures to develop ICTs (Zheng 2008). China officially came online in 1994 with only 10,000 Internet users. The online population drastically increased to 253 million by June 2008, putting it ahead of the United States as the world's largest Internet market (CNNIC 2008). In addition, China has experienced a remarkably rapid expansion of mobile telephony. In 2014, China's mobile phone users reached 1.3 billion (*News.qq.com* 2015), the largest number of any country in the world. Among them, 557 million use mobile phones to connect the Internet (CNNIC 2015). The rapid development of ICTs, particularly the Internet, has accelerated the digitization of traditional media and the convergence of China's media landscape. It has also facilitated the participation of ordinary Chinese people in the nation's political life. This great digital revolution has thus transformed Party-market dominated media events in several ways.

First, the new ICTs have facilitated the free flow of information and created interactive and networked communication platforms for people to participate in media events. Media events are no longer dominated by mainstream media, but are co-represented and re-represented by millions of new media users on multiple digital sites, such as Bulletin Board System (BBS), online chatting groups, and social networking sites (SNS). The extended temporal-spatial mediation of media events in the digital era has made Chinese media events a form of "thickened mediated communication" (Hepp and Couldry 2010).

Second, the development of ICTs has generated various practices of online activism. The alternative and activist use of the Internet has enabled people to communicate beyond the pre-planned agendas set by the Party and the market. Media events are no longer related to just one power centre under "power-money" hegemony; they have become "power-related articulations" (Hepp and Couldry 2010), through which multiple social actors can establish different discursive positions, pursue different agendas and realize different political purposes.

ICT-enabled citizen engagement has challenged China's Party-market dominated media events by taking in voices and actions from civil society. Passive spectators in the pre-reform and reform eras have become interconnected and proactive participants in the new media era (Yu 2006). They can establish networks, exchange ideas, give comments online and talk back to media event producers and organizers. More radically, they can use the Internet to mobilize and organize collective actions online or offline, producing news events within the context of media events to hijack critical moments. The rise in online activism has facilitated informal dialogue between the Party-state and China's civil society, creating deliberative media events in the Internet age.

To describe media events in China's new media era as "deliberative" is not to imply that the government, media and society will embark on rational-critical dialogue with each other within the framework of "deliberative democracy" (Cohen 1989; Habermas 1989). In China, where genuine electoral democracy has not been fulfilled yet, and people's conscious awareness of citizenship and democracy is relatively low, deliberative democracy is almost an "unrealized utopian ideal" (Fraser 1990). Therefore, I adopt Dahlgren's weak and plural view of deliberation to understand Chinese people's political engagement aided by the Internet in media events. Dahlgren prescribes:

> More popular forms of deliberation are needed, and should be spread out as far as possible within citizenry, beyond the formal decision-making centres, into the public sphere and into as many associations and networks of civil society as is feasible. (2009: 89)

Following his argument, the term "deliberative" media events highlights the popular forms of deliberation taking place outside formal political institutions, including the practices of online activism, which have increasingly intervened in media events. Online activism can stimulate popular discussion, improve citizens' political participation and promote interaction between the Party-state and society, having the potential to loosen the dominant forces that control media events in China.

As discussed above, the development of the Chinese media since 1949 has seen China's media events transform from being authoritarian, to marketized authoritarian, to deliberative. The characteristics of China's media events in the three historical periods mentioned previously are summarized in Table 1.[2] This book focuses on the third historical phase of Chinese media events—deliberative media events. It demonstrates that the practices of online activism of ordinary Chinese people have caused a deliberative turn in China's media events.

Significance of the research

This book contributes to studies of the Internet and digital communication in China. It is among the first to use a "media events" framework to examine China's online activism and Internet politics. This framework enables me to study China's online activism in specific socio-historical contexts, and explore its spatial, temporal and relational dimensions with other social agents. The contextualized analysis of online activism allows me to examine China's Internet politics on three interrelated levels: the institutional (regu-

Table 1 Transformation of Chinese media events post-1949

Periods	*Types*	*Media system*	*Media involved*	*Audiences*	*Characteristics*
Pre-reform era	Authoritarian media events	Totalitarian system	Party-controlled official media	Passive spectator	Commemorative, ideology-laden, monologic
Media reform era	Marketized-authoritarian media events	Dual-track system	Party-market controlled official media	Spectator and consumer	Diversified, propaganda-oriented, market-oriented
New media era	Deliberative media events	Dual-track system with rapid ICT development	Party-market controlled official media and Internet-based unofficial media	Proactive participant	Contested, interactive, deliberative

lation and censorship), discursive (symbolic representations) and individual (everyday communicative practices).

This research also contributes to studies of China's media events by considering online activism. Research in English and Chinese on China's media events has hitherto revolved around three typical perspectives: the "ritual critique" perspective, the "Sino-Western" perspective and the "new media events" perspective. The ritual critique perspective focuses on ritualized media events and criticizes their discursive and ideological hegemony, including studies of the CCTV Spring Festival Gala (Lü 2003a, 2009; Pan 2010a; Sun 2007; B. Zhao 1998), China's new millennium celebration (Yu 2009), and the state media's reporting on the 2008 Sichuan earthquake (Sun 2010; Yin and Wang 2010). However, this perspective fails to look beyond dominant media power to examine discursive peripheries in media events. In contrast, the case studies in this book address the key gap in approaching media events as contested platforms. I retain the critical ethos of ritual critique, but shift the focus from hegemonic narratives to counter-narratives. I explore the assorted discursive positions of multiple social actors and their interrelations within and beyond media events.

My research also revises the "Sino-Western" perspective, which underscores oppositional representations between Chinese and Western media on Chinese media events with global impact, such as Hong Kong's handover in 1997 (Lee et al. 2002) and the 2008 Beijing Olympic Games (Price and Dayan 2008). Compared with the ritual critique approach, this perspective takes discursive contestation into consideration. However, this "China-versus-West" model is still framed within the now-outdated and discredited Cold War ideology. Competing political and ideological stands between the socialist regime and Western democrats pre-determine their discursive contestation in representing China's media events. This study moves beyond the Sino-Western perspective by examining China's media events from within. Domestic politics, rather than international politics and relations, is emphasized. Discursive contestation is still the key feature, but it occurs among the multitude of social agents in Chinese society, not between China and other societies. Taking this approach allows the complex interrelations between the state and the non-state to unfold.

I also want to amend the "new media events" perspective. Jack Qiu (2009) proposes the concept of "new media events" [*xinmeiti shijian* 新媒体事件] to characterize China's increasing Internet incidents, in which marginalized social groups actively engage with controversial social issues by making use of new media technologies. Though Qiu takes into account the societal powers and new media technologies in studying the emerging form of media events in China, he loosely uses the classical concept of media events but does not refer to the ritual view contained within Dayan and Katz's original concept. As Dayan commented in an interview with Qiu: "New performances of media events could not deny the classic paradigm of media events . . . the consensus media events still exist, however, their performances evolve . . . we could see both consensus and conflicts in an unfolding event" (quoted in Qiu and Chan 2011: 8). Considering China's media events in the twenty-first century, Wanning Sun makes a similar argument that "the scenario we face is not the replacement of the enhancement of the broadcasting era with the cynicism of the post-broadcasting era, but rather their uneasy coexistence and curious codependence" (2014: 457).

I also discuss societal powers and new media. However, they are strategically situated in the context of media events. What Qiu calls "new media events" are seen here as a consequence of interventional practices of online activism contained by and staged in mega media events. In this way, I acknowledge the validity of traditional paradigms of media events in the Chinese context, but also take into account the new interventional force—online activism—to update them.

Methodologies and approaches

This book uses an integrated qualitative research methodology to collect and analyze research data, including case studies, participant observation, interviews and textual analysis. These micro techniques and methods allow for the incorporation of data across a whole range of mediums (old and new media) and spaces (online and offline). They also facilitate the flexible combination of diverse materials and methods and an appreciation of the multidimensional dynamics of media practices in China's media events. Conceptually, the research is guided by two macro approaches: "media-as-practice" and "articulation".

Nick Couldry proposes the former approach, which—as its name suggests—advocates that researchers regard media as practices, not as texts, economy or institutions. They are to observe what people do "in relation to media across a whole range of situations and contexts" and how media practices (production, circulation and consumption) order other social practices (2004: 119, 128). This approach emphasizes how media practices anchor other types of practices on the micro level; it also addresses the social, cultural and political transformations caused by mediatized social actions on the macro level. Following this tack, this book takes media events as open platforms for multiple social actors to contest, negotiate and interact by using their privileged media resources to exercise various media practices. This enables investigation of the multiple layers of interactions between different communicative practices in media events, particularly between state actors and non-state players. This approach connects concrete media practices with other social actions and societal transformations, resonating with the "articulation" perspective (Grossberg 1993; Hall 1980; Pan 2010b).

Articulation is "the form of the connection that can make a unity of two different elements, under certain conditions" (Hall 1986: 53). It can link "this practice to that effect, this text to that meaning, this meaning to that reality, this experience to those politics" (Grossberg 1992: 54). Zhongdang Pan adopts the theory of articulation to understand China's media and societal changes. He claims that media change is one of the most important social changes in China; however, it is also a "part of the articulation of various forces in the social and cultural formation", while also providing "resources, both ideational and material, for the articulation to take place" (2010b: 518). He suggests that researchers in Chinese media studies "foreground media changes to understand the formation of a media system" and also "decentre the media by locating media changes in the process of the social and cultural changes that they both constitute and enable" (2010b: 518). This approach, according to Pan (2010b), could serve three overarching goals, namely, "theory development", "social and cultural criticism" and "practical instigation".

In this research, media events are utilized to contextualize and address these goals. First, in terms of "theory development", Pan argues that changes in China "represent a unique case that cannot be addressed comfortably with the theoretical apparatus that is largely rooted in very different historical experiences of others". He reminds researchers that "most of the theoretical propositions framed in universalistic terms are abstractions from case studies of non-Chinese" (2010b: 518). As discussed earlier, the concepts and theories surrounding media events have mainly originated in liberal and democratic societies in the West. Research concerning non-liberal and undemocratic nations, such as China, is limited. This book thus tests the validity of Western media event theories in the Chinese context and revises and develops them in light of unique Chinese situations.

Second, in terms of "social and cultural criticism", Pan proposes to criticize "the measures and steps for change that depart from such normative principles as equality, justice, and human agency", exposing "the oppression, distortion, and hypocrisy embedded in the rhetoric for changes" (2010b: 518). This book critiques not only the state media's dominant and hegemonic narratives in media events, but also various practices of online activism in the same contexts, thereby revealing the domination, oppression and resistance of multiple discourses in media events.

Third, in terms of "practical instigation", Pan advocates that researchers "address action issues and provide discursive means that could enable alternative actions and alternative voices, especially those voices that are institutionally silenced" (2010b: 518). This resonates with the theme of this book, which focuses on the modes, practices and characteristics of online activism as well as its intervention in media events. By so doing, this book aims to increase the "critical media literacy" (Kellner and Share 2007) and inform potential practitioners.

Chapter outline and case studies

Chapter 1 reviews the concepts, theories and practices of alternative media and examines the historical development of alternative media in China. It argues that the Internet has become the most important alternative media in present-day China. The chapter then studies China's online activism generated through activist use of the Internet. It examines the main themes and characteristics of online activism in Chinese society and further categorizes it into three major modes—culture jamming, citizen journalism and mediated mobilization. The historical and theoretical origins of the three models of online activism in the West and their performances in China are also discussed. The three modes of online activism are located in three different types of Chinese media events, which are investigated in the following chapters.

Chapter 2 discusses *shanzhai* (copycat) media culture in the context of a celebratory media event, the CCTV Spring Festival Gala. It first explores the CCTV Spring Festival Gala as a ritualized media spectacle and critiques its ideology and market monopoly. It then conceptualizes *shanzhai* media culture as a culture jamming practice with *e'gao* (spoofing) spirit and *shanzhai* ethos. The chapter takes Lao Meng's *Shanzhai* Spring Festival Gala as a case study to examine how the *shanzhai* gala intervenes in the CCTV's power-money-dominated Spring Festival celebration. Further, this chapter argues that *shanzhai* media culture is a counter-spectacular media culture, which challenges the monopoly of the Chinese media establishment's spectacle in both ideology and market. By spoofing and copycatting media spectacles, *shanzhai* media culture resists and criticizes the disingenuous neoliberal logic of China's media reform and development through grassroots' creativity and entrepreneurship.

Chapter 3 discusses citizen journalism as alternative crisis communication in a disastrous media event—the 2008 Sichuan earthquake. This chapter first examines the characteristics and turning points of the Chinese media in crisis communication from the Mao era to the post-SARS era. I discuss the conventional model of China's crisis communication, SARS crisis communication reform and the characteristics of the post-SARS crisis communication. I also examine citizen journalistic practices in the aftermath of the 2008 Sichuan earthquake, including eyewitness reporting, online discussion and networking, and independent investigations. Citizen journalism plays a crucial role in alternative crisis communication in times of crisis. It enables agendas overlooked by government (intentionally or unintentionally) to re-enter the public vision. Media coverage of disasters have become sites for the contestation between "top-down" crisis communication dominated by the Party-state and "bottom-up" crisis communication in the form of citizen journalism.

Chapter 4 examines online *weiguan* (virtual crowd gathering) as networked collective action in scandalous media events in China. It first discusses the history, practices, promises and limitations of China's investigative journalism in exposing political scandals and supervising political power. Online *weiguan* is then proposed as an alternative and popular surveillance. Based on cases on three major online *weiguan* platforms—BBS, human flesh search (HFS) engine (where people track an evil individual down with collaborative man power online), and micro-blogs, I argue that online *weiguan* plays an important role in drawing public attention to scandalous events when mainstream investigative journalism is conservative. Online *weiguan* has become an effective strategy for establishing networks, fostering public opinion and generating pressure upon the government to resolve scandalous events in China.

As well as concluding the study, Chapter 5 offers the concept of "Internet interventionism" to explain the intervention of online activism and its influence on state and non-state interrelations in China. It points out that Internet intervention from both bottom-up (online activism) and top-down (Internet-based governance) has greatly increased the mediation of politics in scope and degree, and has promoted Internet-mediated political communication in China. The Internet has thus become a public interventional tool that multiple social actors can make use of to pursue different political agendas. The interplays between the state and the non-state, which are oriented to the Internet, have not only created deliberative Chinese media events, but also facilitated a deliberative turn in Chinese politics.

1

Alternative Media and Online Activism

Chapter 1 theorizes the Internet as an alternative media and presents online activism as political communication. This chapter first discusses alternative media, the platforms and technological settings of online activism, and provides a historical overview of the development of alternative media in China. The Internet has become the most important alternative media in China today, and online activism takes many forms. This chapter then discusses China's online activism by looking at its major themes, researches and characteristics. It further categorizes China's online activism into three major modes: culture jamming, citizen journalism and mediated mobilization.

Understanding alternative media

Alternative media—also known as "radical media" (Downing et al. 2001), "citizens' media" (Rodriguez 2011), "tactical media" (Garcia and Lovink 1997), "activist media" (Waltz 2005) or "autonomous media" (Langlois and Dubois 2005)—play an important role in "media production that challenges, at least implicitly, actual concentrations of media power, whatever form those concentrations may take in different locations" (Couldry and Curran 2003: 7). They "[open] up cracks in the mass-media monolith" (Waltz 2005 viii) and can act as "unofficial opposition to mainstream media" (Kidd 1999: 113).

Alternative media counteract the concentration of media power by challenging the dominant forms of media production, structure, content, distribution and reception (Fuchs 2011: 298). In terms of media production, alternative media provide affordable platforms for people to participate in symbolic production. The passive audiences of the past have become proactive consumers, or "prosumers" (Toffler 1980), and challenge dominant symbolic production, which is controlled by power, money and knowledge. In terms of media structure, alternative media are usually generated by grassroots organizations instead of hierarchical corporations. They are largely non-commercial and non-profit media, financed by donations, public funding or private resources, and thus operate relatively free from corporate and government influence (Fuchs 2010: 179). In terms of content, alternative media usually provide standpoints which oppose mainstream media. The voices of the minority and marginalized groups—which are likely to be excluded, oppressed or misrepresented in mainstream media—come to the fore in alternative media. In terms of distribution, information from alternative media can be accessed in an open way and at no cost to the consumer, and thus adopts revised notions concerning intellectual property, such as "anti-copyright" (Atton 1999). At the level of reception, alternative media privilege involved audiences over the merely informed (Lievrouw 1994). They encourage audiences to reflect critically upon the content of mainstream media and participate in social actions for social change.

Alternative media embrace a wide variety of forms, such as public speech, dance, graffiti, street theatre, as well as the predictable mass media (press, radio, film, television and the Internet) (Downing et al. 2001). The alternative and activist use of the alternative media has generated various media activism practices, challenging the hegemonic policies, priorities and perspectives of the dominant media culture. Development of the Internet and other new media technologies since the 1990s has represented a new era for alternative media (Downing et al. 2001). The Internet represents the newest, most widely discussed, and most significant manifestation of new media (Flew 2005). Its electronic networks link people and information through computers and other digital technologies, which allow for interpersonal communication and information retrieval (DiMaggio et al. 2001). The Internet has helped "facilitate the growth of massive, coordinated digital networks of engaged activities", and allows marginalized social groups to participate in political activities, and "counter those of the more powerful, not least as expressed in the dominant mass media" (Dahlgren 2009: 190). This technology has been applied in a wide range of projects, interventions and networks, which work against media, cultural and political domination.

Alternative media in China: from big-character poster to micro-blog

There has been a rich repertoire of alternative media in the People's Republic of China. The history of alternative media in China can be traced back to big-character posters [*dazibao*大字报] in the 1960s and 1970s. Big-character posters are wall-mounted posters written in large Chinese characters. They are cheap, sometimes anonymous, and easily seen and read in public spaces, making them an effective means of propaganda, critique, debate, and mobilization in political movements, such as the Cultural Revolution (1966–76) and the Democracy Wall Movement (1978–79) (Downing et al. 2001; Sheng 1990).

From 1978 to 1989, non-governmental periodicals [*minjian kanwu* 民间刊物]—run by poets, liberal college students, social activists and public intellectuals—replaced big-character posters and became the main alternative media in China—*Today* [*Jintian* 今天], *China Human Rights* [*Zhongguo renquan* 中国人权], *Beijing Spring* [*Beijing zhiqun* 北京之春], *May Fourth Forum* [*Wusi luntan* 五四论坛], to name a few. There were at least 55 non-governmental periodicals published in Beijing and approximately 127 published elsewhere in the nation from 1978 to 1989 (Gu 2008; Ran 2007; Wen 2009). These radical periodicals criticized current social problems, proposed critical understandings of the Cultural Revolution and Chairman Mao's rule, and supported the government's liberal and democratic reforms. However, the wave of non-governmental periodicals drastically declined after the crackdown on the 1989 student movement, as the CCP closed nearly all of them down and tightened censorship and regulations (Wen 2009). Due to the CCP's strict control of public opinion and heavy-handed suppression of citizen activism after 1989, alternative media practices almost ceased to exist in China from 1989 to 1994.

Since 1994, however, the emergence and development of the Internet has opened up a new era for China's alternative media and media activism. On April 20, 1994, the National Computing and Networking Facility of China (NCFC) opened a 64kbps international dedicated circuit to Internet through Sprint Co. of the United States. Ever since, China had been officially recognized as a country with full functional Internet accessibility. The rapid development of the Internet has caused China's digital revolution and has had a profound impact on Chinese society. According to Yongnian

Zheng, the primary purpose for the Chinese government to develop Internet technology is to promote China's "nation-state building". The Internet is perceived not only as "a symbol of the modernity of the Chinese state", but more importantly, as "one of the core pillars of sustainable economic growth" (Zheng 2008: 18). However, since 1994, the Internet has not only accelerated China's modernization and globalization, and created a burgeoning ICTs industry, but has also served as an alternative media, facilitating civic engagement, political liberalization and democratization.

As discussed in the Introduction, Chinese media were tightly controlled by the Party in the pre-reform era and were dominated by power-money hegemony in the reform era. The voices of the populace regarding policy critiques, rights defence, political reforms and other democratic issues have long been repressed. Increasing social inequality and injustice, along with China's rapid economic growth, require a media channel beyond mainstream media to represent the views of vulnerable individuals and disadvantaged groups. The Internet has quickly become the most popular and important alternative media in China due to its interconnectivity, interactivity, ubiquity, easy access and low cost, although it is still controlled, regulated and censored by Chinese authorities (Dai 2000; MacKinnon 2010; Tsui 2003; Zhang 2006). Various online platforms have been created, such as electronic mailing lists, BBS, instant messaging software, blogs, audio-visual sites and micro-blogs, which have helped people express, connect, and interact. BBS sites and micro-blogs, the most popular online platforms in China, have played a particularly important role in facilitating online discussion, establishing grassroots networks, forming online public sentiment and organizing collective actions, which have effectively enabled people to articulate alternative agendas and impose pressure on the government to improve governance.

Online activism in China: themes and modes

Online activism is also termed "Netactivism" (Schwartz 1996), "Cyberactivism" (McCaughey and Ayers 2003), or "Web activism" (Dartnell 2006). According to Yang, it refers to "contentious activities associated with the use of the Internet and other new communication technologies" (2009a: 3). It enables people to "extend their social networks and interpersonal contacts, produce and share their own 'DIY' information, and resist, 'talk back' to, or otherwise critique and intervene in prevailing social, cultural, economic, and political conditions" (Lievrouw 2011: 19).

China's online activism appeared very soon after the nation officially came online in 1994. It is built on previous forms of popular contention in Chinese society, such as the peasant revolutions, Cultural Revolution and more recent student movements (Yang 2009a). Applying the Internet to traditional rituals of protests has created innovative tactics and new forms of social activism, thus generating various online activism practices with Chinese characteristics.

Themes of China's online activism

Guobin Yang (2009b) categorizes China's online activism into four themes: nationalism, cultural identities, controversial social issues and political actions. Nationalism was the earliest theme of online activism in Chinese society. In September 1996, university students in Beijing used untitled BBS at Peking University (bbs.pku.edu.cn) to organize anti-Japanese demonstrations to defend China's territorial sovereignty over the Diaoyu/Senkaku islands (Chen 1996). In the following years, nationalism also became the most central theme of China's online activism. BBS sites became major platforms for organizing both online and offline nationalistic protests. Defining cases included the protest against violence committed against ethnic Chinese in Indonesia in 1998; the protest against the NATO bombing of the Chinese embassy in Belgrade in 1999; and the US-China spy plane collision in 2001. As the earliest practice and one of the most important types of online activism in China, online nationalism has evoked a great deal of attention from academics in China and overseas. Scholars have examined responses to Sino-foreign events to discuss how China's nationalistic discourses have transformed in the Internet era, and how online nationalism has influenced China's foreign relations, diplomatic policies and the formation of Chinese civil society (Chase and Mulvenon 2002; Qiu 2006; Shen and Breslin 2010; Wu 2007).

Since the beginning of the twenty-first century, China's online activism has increased in frequency, scale, diversity, and influence, with the proliferation of online platforms and the penetration of the Internet into everyday life. Cultural, social and political forms of activism have taken priority over nationalistic discourses online.

Cultural activism involves online expressions of identities, values and lifestyles which challenge the traditional ethics, morality, values and aesthetics in Chinese society (Yang 2009b). In late 2003, Muzimei, a newspaper editor and columnist in Guangzhou, gained fame by putting the details of her sexual life on her personal blog. She quickly became one of the earliest Internet celebrities and the most famous sex blogger in China; her blog attracted heated discussion in print media, BBS, blogs and online chat rooms across China. The controversy surrounding the "Muzimei phenomenon" [*Muzimei xianxiang* 木子美现象] was widely reported by the *New York Times*, *Washington Post* and other international media (*Sydney Morning Herald* 2003), making Muzimei a popular culture event. The Muzimei phenomenon demonstrated that ordinary Chinese citizens could achieve fame not through mainstream media, but via individualistic, opportunistic and narcissistic self-promotion in the online world of user-generated blogs and forums (Roberts 2010). The Internet has provided the grassroots with an ideal platform to represent themselves, and to show personalities and form identities which are not constrained by traditional social and cultural values.

In addition, the Internet is used to challenge and deconstruct established cultural products and images (blockbusters, TV programs and revolutionary heroes). Some proactive Internet users use web-based digital technologies, such as Photoshop and Flash, to remix established cultural products and images in order to make spoofs, generating popular *e'gao* culture. As Chapter 2 discusses, this playful but satirical web-based subculture cultivates ordinary citizens' subjectivities and critical thinking, challenging the CCP's long-term cultural hegemony (Gong and Yang 2010; Meng 2009; Voci 2010).

Online activism about controversial societal issues covers a wide range of critiques concerning official corruption, environmental pollution, social injustice and human rights violations, generating many Internet incidents with national impact (Yang 2009b). The year 2003 marked the rise of the "Internet incident" [*wangluo shijian* 网络事件] in China. Public sentiment generated online forced the government to release information on the SARS epidemic and abolish administrative procedure surrounding custody and repatriation through the case of Sun Zhigang.[1] Chinese online opinion demonstrated its great supervisory power for the first time in history. Thus, 2003 is popularly dubbed "the year of online public opinion" [*wangluo yulun nian* 网络舆论年]. Since 2003, a year has seldom passed without Internet incidents. In these incidents, grassroots netizens effectively set agendas for the mainstream media's subsequent reporting and the government's further investigation to solve controversial social issues. Having examined numerous episodes, scholars argue that the Internet has empowered marginalized individuals and groups to establish networks, generate online public opinion and have a voice, which has in turn contributed to the development of China's civil society, while also promoting the interplay between the state and society (Cao 2010; Jiang 2012, Qiu & Chan 2011; Yang 2009c).

Online activism about political actions refers to radical and oppositional popular protests facilitated by the Internet. It usually addresses sensitive topics such as human rights, political reforms, the 1989 Tiananmen protests, and the issues of Tibet and Taiwan. This type of online activism is often labelled as "extra-legal" by the Ministry of Public Security, since it directly challenges the legitimacy of the CCP's governance. Therefore, it is highly risky and susceptible to crackdowns. The 2011 Chinese Jasmine revolution is one such example. The anonymous call for a Jasmine revolution was made online, first on Boxun.com, which is run by overseas dissidents, and Twitter, which is blocked in China. It appealed for ordinary citizens to take regular Sunday strolls in thirteen major cities across China to express their dissatisfaction at China's insufficient political reform. The call was then posted in China's cyberspaces by savvy Internet users who know how to get around the Great Firewall, and was spread on China's social media sites. In the "revolution", the Internet was used to propagandize, organize and mobilize pro-democracy protest. However, due to its challenging of the CCP's one-party rule, the protest was quickly shut down by authorities and ended in failure. About thirty-five leading Chinese activists were arrested or detained (Pierson 2011; Ramzy 2011a).

The cases and themes discussed here indicate that online activism in China is precarious. Nationalistic online protest, online self-expression (promotion), and online rightful resistance are relatively safe, though still under regulation and control, whereas the Internet-facilitated pro-democracy social movements that have organizational bases, political aims and offline actions are highly risky. As Zheng and Wu (2005) point out, in China, safe and illegal online activism practices are mostly those "voice activities" which do not pose a direct challenge to the state and aim to facilitate interaction between the state and society. My book focuses on such "voice activities" instead of extra-legal protests because they tend to be more effective than radical practices in engendering change in Chinese society.

The examination of the themes of online activism is helpful for understanding its socio-political functions and impact in China's social, cultural and political domains (see Table 2). However, it is necessary to go further to discuss the strategies and techniques of online activism to see how it is practiced.

Modes of China's online activism

Modes of online activism refer to the common strategies and techniques used in online activism across different themes. Mapping the modes of online activism can delineate and clarify the strategies, techniques, characteristics and purposes of each type of online activism. Moreover, this mapping can make online activism practices more recognizable and provide practical guidance for practitioners.

Three considerations are useful in identifying the major modes of online activism in China. First, frequency—how often online activism is practiced by Chinese Internet users; second, influence—the extent to which online activism practices have affected Chinese society, culture and politics, and attracted media, government and international attention; and third, similarity—that is, online activism practices of a certain mode share strategies, techniques, purposes and functions. Based on these considerations, and informed by others' typologies (Atton 2004; Lievrouw 2011, McCaughey and Ayers 2003; Meikle 2002; Waltz 2005), I identify three major modes of online activism in China: culture jamming, citizen journalism and mediated mobilization.

Culture jamming

The history of culture jamming can be traced back to the twentieth-century European artistic avant-gardes and the Situationist movement from the 1950s to the 1970s. As Lasn argues, culture jamming is on a "revolutionary continuum" with anarchists, Dadaists, Situationists, the sixties hippie movement, and early punk rockers, among others (1999: 99). Culture jamming is also "semiological" (Dery 1993) or "meme" (Lasn 1999) warfare, through

Table 2 Themes of China's online activism

Themes	*Purposes*	*Performances*	*CCP's attitudes*	*Representative cases*
Nationalism	Defend national dignity and interests	Online petition and protest	Supportive with necessary control	NATO bombing of the Chinese embassy in 1999 US–China spy plane collision in 2001
Cultural identities	Challenge cultural hegemony	Blogging and microblogging by grassroots cyber- celebrities, *e'gao* culture	Neutral with necessary regulation	Muzimei's sex blog Hu Ge's *e'gao* video (see Chapter 2)
Controversial social issues	Generate online public opinion and set agendas for mainstream media and government	Internet incidents about social injustice, environmental pollution and official corruption	Precarious according to the themes, venues and timing of the Internet incidents.	Sun Zhigang case Hide-and-Seek incident (see Chapter 4)
Political actions	Pro-democracy	Internet-facilitated social movements	Crackdown	China's Jasmine Revolution

which, cultural jammers "intrude on the intruders, investing ads, newscasts and other media artefacts with subversive meanings" (Dery 1993). One of the classic examples of culture jamming is Jonah Peretti's "Nike Media Adventure". Peretti took advantage of a marketing campaign of Nike that allowed customers to order custom-made shoes by adding a word or slogan. Peretti chose the word "sweatshop" to be printed on his shoes. In subsequent email correspondence with Peretti, Nike refused to fill Peretti's order or give any explanation for the refusal. Peretti posted his correspondence with Nike on the Internet after his order was rejected. His post was soon widely circulated online all over the world and reported by many mainstream media. Nike was embarrassed by the adverse publicity (Cammaerts 2007; Lievrouw 2011; Peretti 2001).

As Peretti concludes, culture jamming is a creative strategy, which "turns corporate power against itself by co-opting, hacking, mocking, and re-contextualizing meanings" (2001: 1). It aims to "reverse and transgress the meaning of cultural codes whose primary aim is to persuade us to buy something or be someone" (Jordan 2002: 102). Culture jamming promotes anti-consumerism (Lasn 1999) and can be taken as a form of consumer boycott (Carducci 2006). However, culture jamming is not only a technique for countering "capitalist corporate brand culture", but can also be used as an intervention "in the realm of politics" (Cammaerts 2007: 72). By using humour, mocking, satire and parody, political (culture) jamming subverts the dominant political discourse and develops counter-hegemonic discourses. Culture jamming, either anti-consumerist or anti-political, strives to "hack into the mainstream public sphere, controlled largely by market and state" (Cammaerts 2007: 73).

Culture jamming has three main characteristics. First, it is counter-cultural, subverting mainstream culture to "reveal and criticize its fundamental inequalities, hypocrisies and absurdities" (Lievrouw 2011: 80). It adopts popular cultural forms, such as billboard alterations, cartoons and remixing, to *détourn* ideologies in politics and everyday life. Second, culture jamming is "media-based movement" (Morris 2001: 27). Poster, magazine, radio and television were early platforms for culture jamming. However, the rise in Internet technologies pushed culture jamming online and turned it into a participatory and collaborative practice. The Internet has become the most popular platform upon which to produce, circulate and consume culture jamming projects (Cammaerts 2007;

Jordan 2002; Lievrouw 2011). Third, culture jamming is what Giddens (1991) calls "life politics". As Cammaerts (2007) argues, culture jamming is inherently political for its resistance to commodification and market monopoly. It proves that small interventions can make a big difference and that everyday subcultures can evoke social change (Carducci 2006; Lievrouw 2011).

In China, the history of culture jamming can be traced back to the sense of irony in popular culture in the 1990s, such as Cui Jian's rock-'n'-roll, Wang Shuo's "hooligan" novels and the T-shirt craze (Gong and Yang 2010; Yu 2007a). The playful and ironic use of words in these products was consistent with the basic techniques and *détournment* nature of culture jamming in the West, and thus can be interpreted as the advent of culture jamming in China. With the development of the Internet, cultural irony and parody have been digitized. Since 2006, culture jamming has emerged in China's cyberspace as a key part of popular Internet culture. It has evolved from the original *e'gao* culture to the current *shanzhai* media culture, poking fun at the cultural products produced by industries and professionals. It inherits the countercultural spirit of culture jamming in the West, but adapts it to China's unique social, cultural and political situations.

Citizen journalism

Citizen journalism is defined as "people without professional journalism training [using] the tools of modern technology and the global distribution of the Internet to create, argument or fact-check media on their own or in collaboration with others" (Glazer 2006). It comprises ordinary citizens' publishing of reporting, opinions, and commentaries which challenge the institutional practices of news production, such as editorial gatekeeping and advertisement-based business models (Deuze 2003; Lievrouw 2011; Ryfe and Mensing 2008). Citizen journalism is also known as "grassroots journalism" (Gillmor 2006), "networked journalism" (Deuze 2007), "open-source journalism" (Leonard 1999) or "participatory journalism" (Lievrouw 2011). Each term captures some, but not all, characteristics of this new form of journalism. However, a striking feature common to these concepts is the growing involvement of ordinary people in online news production.

Citizen journalism first became noticeable during the 9/11 tragedy in 2001 and was became a prominent phenomenon following the South Asian Tsunami in 2004 (Outing 2005). By using

Web technologies ordinary citizens can generate self-styled reporting, which can contain first-person accounts, mobile or digital camera snapshots and video footage. The user-generated content is then posted online through blogs, personal webpages and social media and thus can reach large audiences, just like professional journalism.

In Western societies, the rise in citizen journalism has been due to the crisis in the news industry and the economic and technological challenges it faces (Lievrouw 2011). Citizen journalism challenges traditional journalism as "expert culture and commodity" (Atton 2004: 60) and "seeks to critique and reform the press as an institution" (Lievrouw 2011: 143). Research on citizen journalism in the West has focused on ordinary citizens' crisis reporting in disasters, such as 9/11, the South Asian tsunami, the Iraq War, the London subway bombings, and Hurricane Katrina (Allan 2007; Gillette et al. 2007; Outing 2005; Wall 2009), or on independent media projects such as the Indymedia movement (Atton 2003; Kidd 2003; Morris 2004; Pickard 2006a; 2006b). These empirical studies have demonstrated the transformative potential of citizen journalism and its significant influence on the news industry and political participation.

The rise of citizen journalism in China is attributed to the government's tight control of news production. The mainstream media's investigative reporting on controversial social issues is restricted by their relationship with the CCP (see Chapter 4). Such a highly-controlled media system cannot satisfy people's innate "right to know" and "right to speak"; hence the birth of citizen journalism, with its commitment to free speech and democratic expression. The watchdog function and public nature of citizen journalism are more prominent in China than in the West, as it breaks through the government's tight media control, produces alternative news information, and empowers marginalized individuals or groups for self-articulation (Cao 2010; Reese and Dai 2009; Xin 2010; Yu 2009). Citizen journalism in China is mainly practiced as eyewitness reporting, online discussion and independent investigation. It negotiates with and contests mainstream journalism to represent complex social realities, particularly in times of crisis.

Mediated mobilization

Mediated mobilization is a mode that "relates to the domain of political/cultural organizing and social movements" (Lievrouw

2011: 25). Since the 1990s, the Internet has been employed as a vehicle for collective action. It has been used in traditional forms of collective action, making it easier to organize, coordinate and mobilize across geographic distances and social and cultural boundaries. It has also created new forms of collective action, such as hacking websites, email bombings, online petitions, and virtual sit-ins. As Van Laer and Aelst (2010) argue, the Internet and other new media technologies have shaped and are shaping repertoires of collective action and social movements. Using digital networking devices such as email lists, weblogs, social media sites and mobile phones significantly reduces the "transaction costs" (Naughton 2001) of organizing, mobilizing and participating in collective action. Moreover, networking and information flow, largely enabled by the Internet, allows collaboration and participation in collective action beyond temporal and spatial constraints, generating new forms of translocal and transnational social movements such as the Zapatista resistance in Mexico in early 1990s (Chadwick 2006), the "Battle in Seattle" in 1999 (Juris 2005), the "Arab Spring" revolutionary wave in North Africa and the Middle East in late 2010 (Cottle 2011) and the global "Occupy Movement" in 2011 (Mirzoeff 2011).

According to Yang, the Internet was first used in collective action in China in the 1989 pro-democracy movement (2009a: 28). Chinese students and overseas scholars used e-mail and newsgroups to raise funds for student protesters, issue statements of support, and organize demonstrations around the world. The Internet has gradually become accessible to average urban consumers in China since 1996. It has been increasingly used to organize, mobilize and participate in collective action such as nationalistic events, rights-defence actions, and pro-democracy protests. However, Internet-facilitated collective action in China is different to that in the West in both performance and function owing to China's unique social, cultural, and political environment, particularly the government's restrictions on freedom of association and assembly. In the West, Internet-supported collective action usually consists of large-scale, mass political campaigns which have clear political aims and corresponding offline actions, whereas in China, Internet-facilitated collective action consists primarily of discursive activities which focus on the resolution of a single controversial issue and involve little offline action (Bivens and Li 2010; Zheng and Wu 2005). Internet-based collective action in China is usually conducted in the form of online *weiguan*, through which numerous Internet users discuss a certain controversial issue in a networked and interactive way, generating strong online public sentiment to influence the offline agendas of the mainstream media and the government.

Summary

Culture jamming (represented by *e'gao* or *shanzhai* practices), citizen journalism and mediated mobilization (represented by online *weiguan*) are the three major modes of online activism in China. They have been widely used in online activism concerning various issues and have intervened in mainstream culture, news production and collective action in Chinese society (see Table 3). Admittedly, these modes do not account for *all* genres of alternative and activist uses of the Internet—for example, hacking. It is not considered to be a major mode of online activism for the purposes of this book because its illegal nature and high technological requirements entail that it is practiced relatively infrequently, and it has a less constructive focus than the three modes of activism discussed here.

In addition, the three major modes of online activism are not mutually exclusive and usually co-exist in media events. For example, in some Internet-facilitated collective action which falls into the mode of mediated mobilization, culture jamming and citizen journalism are simultaneously used as important strategies to provide evidence, comments and critiques, and to mobilize online public opinion. Therefore, these modes should be viewed as complementary and mutually constitutive rather than isolated and

Table 3 Major modes of online activism in China

Modes	*Performances*	*Purpose*
Culture jamming	Digitized satires, parodies, spoofs and copycats, as in *e'gao* and *shanzhai* practices	Challenge the mainstream culture in a non-serious, parodic and playful way
Citizen journalism	Eyewitness reporting, online discussion, independent investigation	Challenge news production dominated by the Party and market
Mediated mobilization	Online *weiguan*	Organize and mobilize online and offline collective action

discrete. Moreover, each mode of online activism is not static. It evolves with new media technologies, different Internet habits, as well as changing regulations and policies relating to Internet control and censorship in China.

This book proceeds to contextualize the three major modes of online activism in three different types of Chinese media events. This structure does not imply that a certain type of media event contains a single mode of online activism, nor that a specific mode of online activism only exists in a certain type of media event. It is adopted because it facilitates a detailed analysis of the practices, characteristics, and political impacts of each mode of online activism in concrete contexts. More importantly, locating online activism in media events highlights the interrelations between the mainstream media and the alternative Internet, as well as between the state and the non-state. In this way, the interventional function and the transformational power of online activism become clear.

2

Media Celebration: *Shanzhai* Media Culture as Media Intervention

Shanzhai media culture, a culture jamming practice in China's cyberspace, has intervened into and transformed celebratory and ritualized media events. The complicated interplays between two contesting celebrations of the Chinese Spring Festival embody this dynamic. A grassroots celebration, the *Shanzhai* Spring Festival Gala [*Shanzhai chunwan* 山寨春晚], has challenged the Spring Festival Gala run by CCTV (hereafter, CCTV Spring Festival Gala). The *shanzhai* gala has copycatted the format and content of the CCTV gala with a sense of satire, attracted grassroots participation and pursued entertainment for entertainment's sake. It has resisted the ritualized, televised and elite-dominated CCTV gala, which aims to communicate ideological messages to its audience through elaborate entertainment programs. Thus, *shanzhai* media culture, exemplified by the *shanzhai* gala, has challenged those Chinese media spectacles that are subject to Party-market hegemony. The dynamics of interplay between *shanzhai* carnivals and media spectacles has made Chinese media culture more open, contentious and diverse.

The Chinese term *shanzhai* literally translates as "mountain village" or "mountain stronghold". It originally referred to the mountain stockades of regional warlords or bandits, which were against the authorities and outside official jurisdiction (Xi 2009). The *shanzhai* most familiar to Chinese people is "Liang Shan Bo" [梁山泊] from the medieval folk novel *Outlaws of the Marsh* [*Shuihu zhuan* 水浒传]. Its rulers performed outlaw deeds in the name of the people, defying the corrupt imperial governance of the Northern Song Dynasty (960–1127).

The modern adoption of the historical term *shanzhai* has become prevalent since the early 2000s, when it began to spread throughout the country from Shenzhen, China's first special economic zone in Guangdong province. Retaining the spirit of "nonconformity", "heroism", "self-preservation" and "autonomy" from its original meaning (Ho 2010: 1–2), the modern term *shanzhai* refers to "a blurring of commodity and simulacra: cheap copycats, fakes, pirated goods, local versions of globally branded goods, celebrity impersonators, as well as parodies of mainstream and official culture" (Keane and Zhao 2012: 217). From iPhones to MacBooks, from Coca Cola to Michael Jackson, from the "Bird's Nest" Olympic Stadium in Beijing to the New Year's gala, copycat or clone counterparts are widely consumed by Chinese people in everyday life, making *shanzhai* a national economic and cultural phenomenon.

In this chapter, I first critique the discursive and economic hegemony of the CCTV Spring Festival Gala. The Party-market tandem logic of the Chinese media has produced media spectacles which fulfil the Party's ideological thought work [*sixiang gongzuo* 思想工作][1] and pursue maximum market benefits at the same time. I then theorize *shanzhai* media culture as a performative activity of culture jamming with *e'gao* spirit and *shanzhai* ethos. This grassroots culture acts as resistance to the Party-market hegemony of Chinese media spectacles. The *Shanzhai* Spring Festival Gala demonstrates these performances, their characteristics, and both the enabling and limiting aspects of this emerging media culture.

The CCTV Spring Festival Gala as Media Spectacle

Festivals are ritual activities that occur in a shared spatial-temporal occasion, and are created by human communities in order to maintain their identities, beliefs, customs and cultures (Lü 2003a). The Spring Festival, which is also known as Chinese New Year or Lunar New Year, is the most important traditional Chinese festival. It begins on the first day of the first month in the Chinese lunar calendar and ends with the Lantern Festival on the fifteenth day. By celebrating the Spring Festival, people of Han Chinese descent in mainland China, Hong Kong, Macau, Taiwan and other diasporic Chinese communities create an imagined Chinese nation and consolidate their Chinese identities (B. Zhao 1998).

Local operas and theatrical stunts are essential components of the traditional festival celebrations (Lü 2003a), which vary by region.

Lü argues that Chinese local operas are usually carnivals or comedies with reunion endings. The reunion theme of operas and the collective way of viewing them resonate with the "family-gathering" ritual of the Spring Festival (Lü 2003a). With the penetration of the mass media into everyday life, television has largely replaced traditional theatre. Since 1983, CCTV has run its annual Spring Festival gala on the eve of Chinese New Year. The continuous four-to-five-hour live broadcast is comprised of singing, dancing, traditional vernacular operas, language plays (cross-talk and comedy skits) and other variety shows. It offers a grand visual banquet for millions of Chinese families and has become "a new folk" [*xin minsu* 新民俗] custom, combining popular entertainment with traditional ritualistic forms.

According to Bin Zhao, the gala "helps strengthen the family-centralism on the one hand, and to unify families into the 'imagined community' of the Chinese nation on the other" (1998: 56). It brings the "Confucian dream of 'great oneness'" to "an atmospheric and symbolic realization" (1998: 46). Wanning Sun further explores the ideological functions of the gala and argues that the show delivers "strong messages of patriotism and national unity" packaged as "entertainment, fun and family festivity" (2007: 191). She believes the gala is a successful example of the Chinese state's ideological work in the domestic sphere of private citizens, and as such "demonstrates the ingenuity of the Chinese state in reinventing ways of indoctrinating and educating the nation" (Sun 2007: 191).

The 2009 CCTV Spring Festival Gala bore out Sun's argument. Three hero groups were pre-arranged to appear on the gala stage—those who won honor for China in the Beijing Olympic Games; the aerospace heroes who successfully carried out missions on the Shenzhou No.7 flight; and earthquake heroes who took part in rescue and relief work in the aftermath of the Sichuan earthquake. The three most important national events of 2008 were intentionally recalled and subtly integrated into the gala, which aimed to display China's great achievements under the leadership of the CCP, and more importantly, to enhance the sense of patriotism of Chinese audiences.

The CCTV gala has made use of, but extended and transformed, the original "family-gathering" ritual of the Spring Festival into a new ritual infused with "officially sanctioned ideologies" (B. Zhao 1998: 43). It is not simply an entertainment show for the sake of entertainment, but rather has become an important site for the

Party-state's thought work in the reform era. The significance of the gala's ideological function entails that it is closely connected with state power. It is controlled and regulated by several central government institutions, including the State Administration of Radio, Film and Television (SARFT), the Publicity Department and the Ministry of Culture. The government's interventions—mainly in ideological prescription, content censorship, policy, and financial support—ensure that the CCTV gala runs on the politically-correct track, while also legitimizing the gala's dominant discursive power, which can represent the Party-state and nation.

State-sanctioned discursive power has easily converted into economic monopoly. As a national media event, the CCTV gala is the most profitable TV program with the highest audience ratings in China's TV history. According to national telephone surveys conducted by the CRT market research company, from 20:30 to 24:00 on the eve of the Lunar New Year in 2009, 95.6% of national viewers (about 710 million people) watched the gala (*Xinhuanet.com* 2009a). High ratings translate into high earnings. The 2010 CCTV gala earned 650 million yuan in advertising revenue (*Ifeng.com* 2010), which nearly equalled the yearly advertising revenue of some provincial TV stations. The gala's political clout and market monopoly are mutually reinforcing, consolidating the show's thirty-year domination of the Chinese TV industry (Lü 2009).

The gala attracts the attention of national audiences not only during the live broadcasting process, but also before and after the show. Topics of interest range from who will direct the gala to who will host it; from who will perform in the gala to who will not; from the audiences' pre-gala expectations to the audiences' post-gala critiques. These topics become popular news events that are widely covered by traditional and Internet media and consumed by national audiences. In addition to the domestic market, the gala is also beamed into diasporic Chinese homes via satellite and high-speed Internet cable. Since 2005, the CCTV gala has been broadcast worldwide with English, French and Spanish subtitles on CCTV-9 (English channel) and CCTV-16 (French and Spanish channel) (*Xinhuanet.com* 2005), and thereby has become a truly global media spectacle.

Kellner defines media spectacles as "phenomena of media culture that embody contemporary society's basic values, serve to initiate individuals into way of life, and dramatize its controversies and struggles, as well as its models of conflict resolution" (2003: 11). Examples include the commodity spectacle of McDonald's, the political spectacle of the Clinton sex scandal, and the television spectacle of *The X-Files*. These media spectacles are driven by profit and competition, produced and reproduced by the mass media, and consumed by global audiences, contributing to the rise of the "infotainment society" and "spectacle culture" in contemporary America (Kellner 2003). However, media spectacles in China are different from those in the American context, because they are driven not only by market logic as media products, but also controlled by the Party-state as political promotion. The difference is thus largely determined by the dual-track nature of China's media system in the reform era.

The dual-track media system has two layers of meaning. First, the market mechanism is harnessed to stimulate the development of the media industry. Second, mass media are still bound by their political obligations as mouthpieces of the Party-state. This Party-market intertwined logic is a paradox, and is disingenuous to both neoliberal logic and socialist tradition (Yu 2011; Zhao 2008b). Simply put, the government has incorporated neoliberal elements to promote economic development but retains the Party's authoritarian centralized control (Harvey 2005: 120). This logic, adopted by the media industry, has created media products pursuing simultaneous maximization of economic and ideological benefits, such as the CCTV Spring Festival Gala. As Li Changchun, former Propaganda Chief of the CCP, stated, "the more our cultural products conquer the market, the more fortified our ideological front will be, the better the social benefits" (as cited in Zhang 2011: 161).

On the other hand, this same logic has also created active Chinese audiences. With the deepening of media commercialization and marketization, Chinese mass audiences have evolved from passive into active participants (Tong 2011; Yu 2009). They are no longer monolithic, passively reading, watching and listening to media; instead, they have become consumers, who choose media channels and information, offer critical comments, and participate in media content production. Media spectacles produced under the intertwined logic have increasingly provoked criticism from and lost their grip on these discerning viewers.

In relation to the CCTV Spring Festival Gala, some audiences are not satisfied with advertisements embedded in the show. Some complain that the gala neglects audiences from the South who do not speak Mandarin. Some performers in the gala have even exposed the hidden rules of the show, which usually involve bribery and corruption. More radically, in early 2008, five scholars released a New Spring Festival Culture Manifesto [*Xin chunjie wenhua xuanyan* 新春节文化宣言] and initiated an online campaign against the CCTV's gala. They criticized the show for imposing overly political functions on the traditional Spring Festival, and for simplifying the culturally- rich celebration into merely watching television. They called upon Chinese audiences to distance themselves from the gala (*News.qq.com* 2008).

However, the online resistance did not provoke an active response from national audiences. On the contrary, increased criticism and public discussion attracted more attention, and increased the gala's audience. It was not until 2009 that Lao Meng's *Shanzhai* Spring Festival Gala challenged the CCTV's gala for the fist time, making Lao Meng a grassroots hero and his gala a national media event.

From *Shanzhai* Economy to *Shanzhai* Media Culture

China's *shanzhai* economy originated from the mobile phone industry in Shenzhen. As China's first special economic zone, established in 1980, Shenzhen was the earliest area to trial a complete market economy in mainland China. The central and local governments offered fiscal support, such as tax discounts and support for central and local government budgets. These favourable policies promoted capital flexibility in the local economy and have since attracted millions of investors from China and abroad (Zhu and Shi 2010). After three decades of development, Shenzhen has come to represent China's reforms and globalization, as well as becoming a frontier city for risk investment, free trade, entrepreneurship, innovation and the *shanzhai* economy.

The development of the *shanzhai* handset industry started in Shenzhen and other nearby cities in the Pearl River Delta region in 2005. In that year, Taiwan's cell phone chip solution company MediaTek (MTK) launched a new product featuring "all-in-one functionality", which combined the motherboard and software (Tse, Ma, and Huang 2009). MTK's integrated circuit (IC) design solution substantially reduces the Research and Development (R&D) time and cost involved in the traditional handset industry, and enables manufacturers to spend most of their resources on technological service and exterior design. At the same time, it transforms the traditional manufacturing process into "manufacturing networks" (Zhu and Shi 2010: 36–37). In these networks, each contributor to the *shanzhai* mobile phone has a specific focus, such as IC design, hand-model making or final assembly. The *shanzhai* model has greatly challenged the entire supply chain of transnational handset companies, from chip R &D, to manufacturing and sales. It has lowered the bar for entry into the mobile telecommunications industry, and reduced costs and market prices. The rise in *shanzhai* handset companies has also jeopardized the dominance of international mobile phone companies from Europe, the United States and Japan. Their total market share dropped from over 90% in 2000 to under 50% in 2008, whereas the *shanzhai* phone market share increased to over 30% (Kao and Lee 2010). Low-end *shanzhai* mobile phones quickly occupy the market in China's second- and third- tier cities and vast rural areas, providing affordable mobile phones for low-income "information have-less" people (Qiu 2009).

Authorities usually condemn *shanzhai* mobile phones, like other *shanzhai* products, as "piracy", "counterfeit", "copycat" or "imitation" (Leng and Zhang 2011), because most *shanzhai* companies have established their business by copycatting the products of famous brands. Nokia becomes Nokir, Samsung becomes Samsing, and Sony-Ericsson is rendered as Suny-Ericsson. Nevertheless, copycatting is not entirely destructive; it can be an effective marketing strategy for *shanzhai* manufacturers to build and develop brands in the start-up period (Leng and Zhang 2011). When *shanzhai* brands are widely recognized by consumers and firms have grown to a certain size, manufacturers often try to get rid of their initial *shanzhai* colors and enhance their innovativeness and competitiveness (Leng and Zhang 2011). This "imitation-plus-innovation" model has transformed many *shanzhai* manufacturers from imitators into emerging indigenous adaptors and innovators (Zhu and Shi 2010). Many small companies have quickly expanded to compete with national or international brands. For example, mobile phone manufacturer Tianyu started out as a *shanzhai* maker but has now become a leading local brand. In the third quarter of 2008, itsmarket share reached 8.1%, making it China's fourth largest mobile phone manufacturer after the three world-leading brands of Nokia, Samsung and Motorola (*Tech.qq.com* 2009).

The Chinese government officially discourages the *shanzhai* model and tries to control *shanzhai* products on the grounds that they infringe intellectual property rights, disrupt the market and destroy genuine Chinese innovation (Ho 2010). However, the immense material benefits from making *shanzhai* products means that pragmatism frequently outweighs ethical standards (Ho 2010). *Shanzhai* manufacturers believe that imitation is the first step towards genuine innovation and attempt to legitimize the *shanzhai* model as "primary innovation" (Kao and Lee 2010), and thus consistent with President Hu Jintao's pledge to build China into "an innovation-oriented country" (Keane and Zhao 2012). The *shanzhai* model has become a practical and effective path for China's small companies to compete with Western capital in the globalized market economy, and increased the gross domestic product (GDP) of both local and central governments. Therefore, in reality, the government turns a blind eye to *shanzhai* manufacturing.

The rapid development of the *shanzhai* phenomenon has seen the *shanzhai* ethos go beyond its original economic realm and expand into the cultural sphere (Ho 2010; Wu 2006). *Shanzhai* has adapted various performative activities to imitate established mainstream cultural icons with a strong sense of self-affirming play. These performances include *shanzhai* Chairman Mao, *shanzhai* Tiananmen, *shanzhai* Olympic torch relay, *shanzhai* celebrities, and *shanzhai* Spring Festival gala. *Shanzhai* culture retains the main characteristics of the *shanzhai* economy, such as imitation, adaptation and innovation. However, as performative activities, it has much richer discursive and cultural meanings, because it usually challenges the "highly centralized Chinese political context" (Ho 2010) embodied in established cultural products. Moreover, this culture also incorporates key aspects of *e'gao*—a culture jamming practice in China's cyberspace which is playful, parodic and counter-hegemonic.

E'gao translates as "wicked fun". The genre emerged in China's cyberspace in 2006 as a "popular subculture that deconstructs serious themes to entertain people with comedy (*sic.*) effects" (Huang 2006). It covers "all types of audiovisual spoofs, oft-times by taking advantage of the transformative capability of digital technology as well as distribution power of the Internet" (Meng 2009: 52). *E'gao* is "a new technology-enabled cultural intervention" (Gong and Yang 2010: 4). By using Photoshop, Flash and other digital remixing technologies, *e'gao*-makers poke fun at professionally and industrially-produced cultural products, including blockbusters, Red classics (revolutionary films, literature, and heroes), and media celebrities, in a parodic style. Through remixing parts of the originals with other texts, *e'gao* works "create ironic incongruity that triggers humor and laughter" (Gong and Yang 2010: 5). They articulate social criticism and grievances, equipping the public to critically understand social realities in a non-serious, non-conventional, artistic and entertaining way.

Among the plethora of *e'gao* works in China's cyberspace, Hu Ge's spoof video, *The Bloody Case of a Steamed Bun* [*Yige mantou yinfa de xue'an* 一个馒头引发的血案], was the biggest Internet sensation (Gong and Yang 2010; H. Li 2011; Meng 2009; Voci 2010; Yu 2007a). In December 2005, famous film director Chen Kaige released his blockbuster *The Promise* [*Wu ji* 无极]. The film was the most expensive in China's history, with a budget of 300 million yuan; it therefore attracted huge attention from media outlets and the public. However, Chen's film disappointed millions of viewers and evoked overwhelming criticism. Hu Ge, a 31-year-old sound engineer and freelancer based in Shanghai, produced a spoof . He edited scenes from a pirated copy of Chen's film and re-organized them in the format of a popular CCTV program, *The Chinese Law Report* [*Zhongguo fazhi baodao* 中国法制报道]. The love story in Chen's film was remade into a murder case reported by a TV program. The two irrelevant hypotexts were re-edited, remixed and re-dubbed, such that they poked fun at each other with humorous and satirical effect. In January 2006, Hu's spoof became the hottest hit on the Chinese Internet as it circulated on major audiovisual portals, weblogs and BBS forums. The grassroots playful and public critique of the cultural elite soon became a popular culture event and was enthusiastically covered by mainstream media. Hu Ge was dubbed "the father of *e'gao*" [*e'gao zhifu* 恶搞之父], and established his fame and reputation as a cyber hero overnight.

Scholars propose various ways of understanding *e'gao* culture. Meng (2009) argues that *e'gao* is a decentralized form of communication that deconstructs authorship and decentralizes cultural production and distribution, though it is constantly being regulated by the state's recentralizing rules and policies. Voci interprets *e'gao* as a form of participatory culture and enlightenment. She argues that *e'gao* parodies develop a contemplative humor that does not ask its participants to act, but invites them to see and understand realities differently from conventional media (2010: 122). Taking the Internet as the main platform for circulation and consumption, *e'gao* culture is a form of digitized culture jamming practice in China.

In 2008, which was proclaimed by the Chinese mass media to be the "Year of *Shanzhai*" [*Shanzhai nian* 山寨年] (X. Wang 2009), *e'gao* culture started to integrate with the popular *shanzhai* economy, generating *shanzhai* media culture. As a mixed culture, s*hanzhai* media culture broadens the territory of the *shanzhai* economy from the manufacturing industry into cultural production. At the same time, it absorbs the *e'gao* spirit in its discursive practices, producing various *shanzhai* media products.

In terms of content, *shanzhai* media products differ from former *e'gao* works. The *e'gao* works are highly reliant on remixing original media texts, whereas *shanzhai* media products are more innovative and original, with little copy-and-paste editing work. They sometimes look funny and parodic like *e'gao* works, but these effects are usually unintentionally created by low-quality production and their great contrast to the original high-quality production.

In terms of format, *shanzhai* media products use the strategies of genre imitation and appropriation, rather than remixing genres as *e'gao* do. Most *shanzhai* media products simply copycat the flagship programs of CCTV, such as the Spring Festival Gala, *Network News* [*Xinwen lianbo* 新闻联播], and *Lecture Room* [*Baijia jiangtan* 百家讲坛]. The main purpose of the *shanzhai* programs is to challenge CCTV's discursive and commercial monopoly, rather than merely produce parodic effects.

In terms distribution and consumption, both *shanzhai* and *e'gao* are Internet-based, though they are usually produced in offline studios or theatres using professional TV-film recording and editing equipment. They are also widely reported online as cultural items and entertainment news and reposted and commented on by hundreds of thousands of netizens on social media.

S*hanzhai* media culture thus manifests both continuity and discontinuity with *e'gao* culture and the offline *shanzhai* economy, incorporating their essences in a new cultural practice. The most representative and influential case of this mixed culture is *Shanzhai* Spring Festival Gala—a copycat of the CCTV Spring Festival Gala. The rise and fall of this *shanzhai* gala has demonstrated the promises, limitations and evolution of the *shanzhai* media culture.

Shanzhai Spring Festival Gala: Counter-Spectacle Entrepreneurship

> Hello everyone, Happy Spring Festival...The year 2008 has flown away. The year 2009 has already come. No matter how serious the global economic crises are, ordinary people's lives are still the most important. If you want to cry, just cry out; if you want to laugh, just laugh out loud. Pour all your complaints out from the heart, because you will not be able to afford expensive medicines if you become depressed or sick . . . If you buy a car, don't drive it so often. The increasing petrol price is unaffordable . . . May ordinary people's lives become better and better. (*Shanzhai* Spring Festival Gala, January 28, 2009: opening remarks)

Lao Meng, whose full name is Shi Mengqi, was born in 1972. He is an IT worker, event planner and producer of wedding videos. In 2002, Lao Meng moved to Beijing from Southwest China. As a migrant, he often spent the Spring Festival alone, and so watching the CCTV Spring Festival Gala had become a habit on New Year's Eve. For him, the gala was an extravaganza hosted by a national TV station, performed by famous stars and sponsored by big corporations, leaving few opportunities for everyday people to participate. Therefore, he decided to make a grassroots gala to entertain rural migrant workers and students who could not return home for family gatherings during the Spring Festival (*People's Daily* 2010). Against the backdrop of the popularity of *e'gao* culture and the *shanzhai* economy, Lao Meng started to prepare his *shanzhai* gala at the end of 2008 with his friends. He wanted to make an unpolished and relaxed grassroots show with a little bit of satirical fun.

On November 23, 2008, Lao Meng posted a "call for programs" on his blog. He addressed his compatriot netizens as follows:

> The purpose of running this gala is to entertain the people for the sake of entertainment. Though we are not as rich as the CCTV, we can collect the best creative ideas and the most excellent programs from the nation's people. All friends with talents are welcome to join us. This will be a Spring Festival Gala for us ordinary people. (Lao Meng's blog, November 23, 2008)

In this blog entry, the slogan of his *shanzhai* gala was proposed: "A People's Gala Held by the People, for the People" [*renmin chunwan renmin ban, banhao chunwan wei renmin* 人民春晚人民办，办好春晚为人民]. This slogan naturally reminded people of one of the best-known Maoist slogans, "serve the people" [*wei renmin fuwu* 为人民服务], particularly because it was written in big Chinese characters that intentionally imitated Chairman Mao's handwriting. Lao Meng echoed Mao's aspiration that the people would become the real masters of the country, though of course his vision was otherwise entirely different to Mao's. On the car he drove across Beijing to promote the gala, Lao Meng printed materials the words "Challenge CCTV and wish all Chinese a happy Spring Festival" [*Jiaoban yangshi, gei quanguo renmin bainian* 叫板央视，给全国人民拜年]. He created a CCSTV logo by inserting the letter 'S' into CCTV and had it printed on microphones, banners and cars. He explained that CCSTV was short for China Countryside Television [*Zhongguo shanzhai dianshitai* 中国山寨电视台], and had nothing to do with CCTV. However, audiences who were used to seeing the CCTV logo found CCSTV unmistakably *e'gao* in tone.

Lao Meng's spirited defiance immediately caught public attention. On November 29, 2008, *Beijing Times* first reported Lao Meng's gala under the headline "Beijinger to run *shanzhai* version of Spring Festival Gala, challenge CCTV" [*Shimin yuban shanzhaiban chunwan jiaoban yangshi* 市民欲办山寨版春晚叫板央视] (Jiang 2009). With the Spring Festival just two months away, people were eager to witness the fight between 'ant' and 'elephant'. Many traditional media and Internet media interviewed Lao Meng and reported on his gala plan, and he soon became a media celebrity for his open challenge to the CCTV.

As an IT worker, Lao Meng understood the Internet, and used it to promote, organize and broadcast his gala. He established an official portal site (www.ccstv.net) to promote the gala and to post updates. Through the official website, Lao Meng could interact with netizens and understand the real needs of ordinary audiences. Netizens could recommend their favourite grassroots stars to perform in the gala, make comments and offer suggestions. After the announcement of Lao Meng's show, more than 300 people expressed interest to be volunteers for the show. About 300 original programs were recommended by netizens. Six enterprises also wanted to sponsor his gala (*Sohu.com* 2008).

The *shanzhai* gala not only caught the traditional media's attention as a news event, but also attracted its participation and cooperation. Guizhou Satellite Television, a provincial satellite TV channel with a signal capable of reaching all regions in China, intended to cooperate with Lao Meng and provide a live broadcasting platform. The two parties quickly signed a contract for mutual benefit. For the *shanzhai* gala, a satellite TV channel could provide a platform with equal reach to that of the CCTV gala. The potential audience for the *shanzhai* gala was no longer limited to Internet users and could now include millions of TV viewers. Guizhou Satellite Television is based in the less-developed region of western China and had advertising revenue of only 500 million yuan in 2008—less than the CCTV gala's revenue for one night (*Baoye.net* 2009). Broadcasting the highly popular *shanzhai* gala presented the Guizhou station with an opportunity to increase its national market share. The support from the local TV station and the high degree of attention from the cyberspace community meant that the *shanzhai* gala became a new rival for the CCTV gala.

When the dominant position of the CCTV gala was threatened, the State Administration of Radio, Film and Television (SARFT) and other related governmental departments intervened through administrative means to protect the interests of the national broadcaster. In late December 2008, SARFT issued an informal notice ordering all television stations not to participate in, broadcast or report the *shanzhai* gala (Jiang 2009). Under SARFT's pressure, Guizhou Satellite Television backed off at the last minute. The contract between the *shanzhai* gala and Guizhou Satellite Television thus became invalid overnight, and the CCTV gala no longer faced any competition from television. This was similar to the fate of Hunan Satellite TV's reality show, *Super Girls* [*Chaoji nüsheng* 超级女声], in 2005. When the show's high national ratings overtook CCTV's *Network News*—the most-watched news program in China—SARFT immediately issued orders to regulate reality shows on local television stations by adjusting their broadcasting times and setting restrictions on length, content, participants and judges (*Cohiba.blogcn.com* 2005). This aimed to crack down on local rivals and maintain CCTV's dominant status in China's television industry. The actions illustrate the irreconcilable tensions between China's central and local media groups. On the one hand, the government encourages local media groups to become bigger and stronger through market competition; on the other hand, the media market is not transparent and heavily state-controlled. Thus, market principles can easily be violated by the government's administrative orders, and any challenge to the state's media monopoly can be stopped.

In addition to the partnership with Guizhou Satellite Television falling through, the portal website that had promised to broadcast Lao Meng's show live suddenly ceased to cooperate with him (*Ido.3mt.com.cn* 2009). The *shanzhai* gala had therefore lost its most important broadcasting platform. The government also warned Lao Meng that his public gala did not have "performance approval" from the Bureau of Cultural Affairs (Lin 2011). The venue booking for the show was subsequently cancelled by the provider without explanation.

Under pressure from the SARFT and other governmental bureaus, Lao Meng was forced to give up his original plan and look for new allies. Finally, Macau Asia Satellite Television (MASTV) agreed to broadcast the *shanzhai* gala live. As a private media company in the Macau Special Administrative Region, MASTV was free from SARFT regulations. However, the signal of private TV stations in China's Special Administrative Regions, such as Hong Kong's Phoenix Satellite Television and Macau's MASTV, only reaches star-rated hotels, news agencies and research institutions. Without a national broadcasting platform, the *shanzhai* gala would be unable to compete with the CCTV gala. In order to avoid government intervention, Lao Meng relinquished the idea of broadcasting live and instead held the gala, publicized as a "rehearsal", in a hot springs resort in suburban Beijing on January 22. The "rehearsal" was recorded as the formal show, and broadcast by MASTV a few days later on Chinese New Year's Eve (Jiang and Ma 2009), but only a very limited audience watched it.

Having failed to broadcast the show live via television and the Internet, Lao Meng and his gala group intended to put the recording online so as to enable a greater audience to view the show. However, they could not upload the videos to nearly sixty Chinese video-hosting websites as long as the clips were tagged as "*shanzhai* chunwan" (Lam 2009). Searches for "*shanzhai* chunwan" on major audio-visual content websites resulted in the message: "the content you are searching for may be related to illegal issues according to laws, principles and policies, and therefore could not be founded" (Jiang and Ma 2009). In an interview, Lao Meng said the biggest difficulty in preparing the gala was not financial or technological problems; instead, it was some "invisible resistance force" from above (Jiang and Ma 2009). Just as Lao Meng expected, major audio-visual websites received a notice from the China Internet Audio-Visual Program Service and Self-Discipline Alliance

(CIAPSSA) around January 20, 2009, ordering all members to boycott the *shanzhai* gala.

CIAPSSA was established by several national Internet media organs, such as *CCTV.com*, *People.com.cn* and *Xinhuanet.com*, in February 2008. By July 2008, more than 150 websites that provided audio-visual content services had joined the alliance. However, CIAPSSA is not simply a business organization, but a government-supported association jointly led by SARFT and the Ministry of Information Industry. The television industry and the Internet audio-visual industry in China are not mutually independent, but intertwined through SARFT's administrative power. Any attempt to challenge SARFT-supported television content through alternative Internet channels is doomed to failure, due to SARFT's strategic regulation under the guise of CIAPSSA's self-monitoring, as well as the conventional self-censorship of China's Internet. This could well explain why Lao Meng's Internet allies abandoned the otherwise mutually beneficial live broadcasting plan at the last minute, and why video clips of the *shanzhai* gala were not uploaded online.

Although the *shanzhai* gala could not realize its aim of rivalling the CCTV gala, it became an important cultural icon, pushing China's *shanzhai* media culture to a climax. It sparked an entire genre of grassroots and Internet-based copycat galas, such as the College Students' Spring Festival Gala, South China Spring Festival Gala, Senior People's Spring Festival Gala and Migrant Workers' Spring Festival Gala, as well as other *shanzhai* programs that imitate the CCTV brand. Lao Meng's *shanzhai* gala also expanded the discussion of *shanzhai* from the economic to the cultural sphere. The previous debate about the *shanzhai* economy which took place among private entrepreneurs, lawyers and policy-makers developed into a national debate on *shanzhai* culture more broadly. In intellectual circles, the main debate is whether *shanzhai* culture is innovation or piracy. Some people hold very positive attitudes to *shanzhai* culture, believing that it embodies the wisdom and creativity of the grassroots. They argue that *shanzhai* culture has promoted multicultural development and has created a more open and dynamic cultural environment. As art critic Xie Yuxi has commented:

> Cultural resources are distributed and occupied unequally and unfairly in China. A small group of people dominate cultural

> production and manipulate the Chinese people's habits of cultural consumption. The emergence of *shanzhai* culture marks marginalized culture's challenge to the mainstream culture. The people are not satisfied with being passive culture-consumers. They are starting to participate in cultural production and show their subjectivity and creativity. (*Beijingreview.com.cn* 2009)

Conversely, some people criticize *shanzhai* culture as piracy and infringement, arguing that *shanzhai* culture disobeys the ethics of the market economy and impinges on intellectual property rights. Professor Ge Jianxiong at Fudan University has argued: "if we tolerate *shanzhai*, real cultural innovation will never be realized" (*Beijingreview.com.cn* 2009). Ni Ping, a well-known anchorwoman from CCTV and a member of the Chinese People's Political Consultative Conference (CPPCC), submitted a proposal at the CPPCC in March 2009 calling for the legal prohibition of *shanzhai* culture in order to support original cultural works (X. Wang 2009). Ni's anti-*shanzhai* proposal immediately sparked debate on the Internet. Most netizens criticized her elite and conservative view on cultural production and supported *shanzhai* culture, which they believed represented the grassroots' spirit and people's real needs. Ni's proposal was not adopted by the CPPCC. The government elected not to respond to the rise of *shanzhai* culture with words, but with actions.

On January 5, 2009, seven government institutions, including the SARFT, the Ministry of Public Security and the State Council Information Office, launched a crackdown on "the wave of online smut" [*hulianwang disu zhifeng* 互联网低俗之风]. The anti-smut campaign aimed to clean up vulgar content online, which was labeled by the government as "immoral" and "tasteless". Such content was accused of violating public morality and harming the physical and mental health of young people (Lam 2009), and these institutions proposed deleting content and shutting down websites. Much *shanzhai* and *e'gao* content became targeted in the campaign due to the political implications of its banter; Rebecca MacKinnon commented that "the technology used to censor porn has ended up being used more vigorously to censor political content than smut" (as cited in Lam 2009).

Though the government did not formally ban *shanzhai* cultural production, it intensified Internet censorship to control the distribution and circulation of *shanzhai* cultural products, which led to the decline of *shanzhai* culture. The government imposes more regulation and control on *shanzhai* cultural products than on the offline *shanzhai* economy, because they challenge the state-sanctioned mainstream ideology and cultural values to a greater extent.

Meanwhile, Lao Meng adjusted his strategies to continue his gala dream. Learning lessons from the 2009 gala, he abandoned his ambition of challenging the CCTV and changed the name of his show to "Folk Spring Festival Gala" [*Minjian chunwan* 民间春晚] in 2010. Compared with the original title of *Shanzhai* Spring Festival Gala, which was full of cynical spirit, the folk gala was a neutral title with little grassroots or revolutionary ethos. In promoting it, Lao Meng gave up the radical slogan against CCTV; instead, he proposed to promote China's folk culture. With this moderation, his folk gala smoothly obtained performance approval from the Bureau of Cultural Affairs, and so it could be broadcast live.

Viewing the folk gala online (*Minjian chunwan* 2010), its commercial nature is striking. In the first five minutes of the show, about ten enterprises are promoted as their representatives greet the audience with "Happy Spring Festival", which was a conventional way for the CCTV gala to embed advertisements. Lao Meng's adaptation to the government's regulations and the commercialization of his show resolved the earlier contention between the *shanzhai* gala and the CCTV gala, ensuring the safety and profit of his 2010 gala. In 2011, Lao Meng prepared to run his third gala. However, due to a business dispute with his gala partner, Lao Meng became a defendant in a legal case, and the 2011 gala did not come to fruition. In November 2011, Lao Meng stated that he would not run any form of Spring Festival gala in the future, but would instead prepare a Valentine's Day Gala in 2012 (*Ent.qq.com* 2011). After three years of entrepreneurship in running his grassroots Spring Festival gala, Lao Meng left this controversial field altogether and started to explore new business opportunities.

Lao Meng's *shanzhai* gala accepted the authorities' amnesty and was "bought off" by capital. His 2010 gala no longer represented the grassroots spirit and culture, but instead had been turned into a real commercial show. The celebration of the Spring Festival had thus become a "cash cow" for not only the CCTV but also the show that had once opposed it (*Xinhuanet.com* 2010). This phenomenon resonated with Lin's argument that "once popular culture is commoditized . . . its potential to be a liberating form of expression is lost" (2011: 62). The power-money hegemony, which protected the longevity of the CCTV's media spectacles, had also become the main force for restraining the counter-spectacle. Trapped between government and market forces, the *shanzhai* gala was not able to keep its revolutionary spirit and independence, and ultimately had to compromise.

The *shanzhai* gala's surrender can also be explained by the theory of "incorporation" (Hebdige 1979). Incorporation refers to the process through which a subculture or a tactic that is different from the dominant one is naturalized and absorbed by mainstream hegemonic culture, deprived of its force as an oppositional body, and integrated into the existing social cultural order (as cited in Dai 2007). Hebdige (1979) further points out that there are two forms of incorporation: commodity and ideology. The evolution of Lao Meng's *shanzhai* gala has shown both forms of incorporation. The commodity form of incorporation was demonstrated through the commercialization of his 2010 gala, and transformed the independent *shanzhai* gala with public service nature into a profitable show. The ideological form of incorporation was evident in the government's various regulations and censorship around the *shanzhai* gala and its online distribution, and weakened the cynical spirit of the gala and pulled it back on to the "politically correct" track. Lao Meng, who established his reputation as a grassroots hero by challenging the domination of the CCTV, has turned into a successful entrepreneur in commercial gala planning. Similarly, Hu Ge, the father of the *e'gao* culture, told me he was not interested in making *e'gao* works any longer. He planned to turn to the big screen and become a film director; some film companies contacted him with a view to investing in commercial comedy films directed by him (Personal communication, July 22, 2010). The two representative cultural activists, of *e'gao* culture and *shanzhai* media culture respectively, ultimately abandoned their attempts to confront the established media centre from the periphery, and instead were incorporated by the power-money hegemony.

Summary

Shanzhai galas are grassroots media carnivals that intervene in and challenge the Party-market dominated media spectacles. According to Bakhtin, carnivals "celebrated temporary liberation from the prevailing truth and from the established" and "marked the suspension of all hierarchical rank, privileges, norms, and prohibitions" (1984: 10). When audiences were fed up with the CCTV gala, which serves the Party and the market in the name of entertainment, they started to produce their own carnivalesque galas to serve people's varied entertainment needs. Bakhtin also argues that "carnival is not a spectacle seen by the people; they live in it, and everyone participates because its very idea embraces all the people" (1984: 7). The *shanzhai* galas recruit creative ideas from the grassroots and are open to all people, particularly the lower social strata of China's post-reform era, such as migrants, laid-off workers and retirees.

Moreover, the great contrast between *shanzhai* and CCTV galas in discursive styles and themes, despite their similarity in genre, creates "grotesque imagery" (Bakhtin, 1984). The central principle of grotesque imagery, according to Bakhtin, is "degradation", which aims to lower "all that is high, spiritual, ideal, abstract" to "the material level, to the sphere of earth and body in their indissoluble unity" (1984: 19–20). By deconstructing the myth of the official celebration, *shanzhai* carnivals have degraded the privileged status of the CCTV gala, and have shown that everyone can make a cultural choice about entertainment. The parodic effects and laughter, caused by either deliberate or accidental degradation in the *shanzhai* galas, have produced "free and critical historical consciousness" (Bakhtin 1984: 73), which has the potential to overthrow the official ideologies embedded in the CCTV spectacle.

The rise of *shanzhai* media culture is an inevitable result of the Party-market intertwined logic of China's media reform and media development, and also simultaneously a sincere form of rebellion against this logic from the bottom up. It contests the power-money manipulated media centre from periphery, with tactical and strategic actions akin to a guerrilla battle. Though guerrillas sometimes fail and their effect is not dramatic, they frequently open up a space for negotiation, interaction, and mutual constitution between the dominant media power and the interventional media force. As a result, CCTV has absorbed major characteristics of its *shanzhai* competitors, such as grassroots participation, online broadcasting and interaction, and public service orientation. While these changes have been limited, CCTV has reformed its media spectacles to maintain the station's dominant status.

Accordingly, the CCTV Spring Festival Gala has started to expand its broadcasting channels from television to the Internet and mobile media. In 2010, over 78.5 million people watched the CCTV gala via China Network Television (CNTV)—CCTV's official online broadcasting platform established in December 2009—as well as other portal sites. Approximately 8.21 million domestic viewers watched the show via CNTV's mobile phone service, and 3.87 million overseas viewers watched it via iPhone (*News.cntv.cn* 2010). CNTV also collaborated with fifteen other websites to organize an interactive activity, "Wish the Motherland a Happy Spring Festival" [*gei zuguo muqin baidanian* 给祖国母亲拜大年], during the four-hour gala. Over 300 million people took part in the activity online (*CCTV.com* 2010). Moreover, the CCTV gala has started to invite more grassroots talent to perform in order to increase its popular appeal. Since 2011, CCTV and CNTV have produced a reality TV show called "I Want to Perform on the Spring Festival Gala" [*Woyao shang chunwan* 我要上春晚]. Grassroots artists are encouraged to upload videos of their performance online. The applicant whose video secures the most online votes wins the opportunity to perform on the CCTV gala; migrant worker duo "Rising Sun and Masculinity" [*Xuri yanggang* 旭日阳刚] and subway singer "Xidan Girl" [*Xidan nühai* 西单女孩] performed onstage at the 2011 CCTV gala with the support of millions of netizens. In addition, the CCTV gala has started to weaken its commercial nature and enhance its orientation towards public service. In 2012, the gala rejected any form of advertising for the first time, and aimed to please the people (*Venturedata.org* 2013).

The complex interplays between the official media spectacle and *shanzhai* media carnival have thus made the ritual of celebrating Spring Festival more contentious. The conventional celebratory media event dominated by the state media has become what Kellner (2003) calls a "contested terrain", in which different social forces compete to represent different interests and agendas through their own forms of celebration. In this process, spectacular and counter-spectacular media powers shape and constitute each other to create a more open, dynamic and diverse Chinese media culture.

3

Media Disaster: Citizen Journalism as Alternative Crisis Communication

Disaster is a state in which the social fabric is disrupted (Fritz 1961) or a "destructive episode" that involves death and damage (Boin 2005). It is "objective geographical process(es)" (Hewitt 1983: 5) and a "recurrent feature of human life" (Alexander 2005: 25). In times of major disasters, mainstream media usually interrupt the regular broadcasting schedule to report gory details from the scene, update the number of dead and missing, and cover rescue and relief work. Because the broadcast continues non-stop for hours or even days, the national or international media events it produces are vividly named "disaster marathons" (Katz and Liebes 2007). Representative events include the tsunami in Southeast Asia in 2004, Hurricane Katrina in 2005 and Japan's 2011 earthquake.

The present chapter investigates the evolving interrelations and complicated interplays between the government, media and public in media events involving disasters in China. I argue that citizen journalism—a major mode of China's online activism—has become an alternative way of doing crisis communication from the bottom up, and has challenged the conventional representation and management of crises. The chapter is structured in two main sections. The first focuses on turning points in China's crisis communication post-1949—specifically, the 2003 SARS epidemic and the 2008 Sichuan earthquake. It examines the characteristics of crisis communication in China with particular attention to the changing role of media in three historical phases, which are divided by the two turning points. The second section of this chapter studies three forms of citizen journalistic practices in the 2008 Sichuan earthquake—eyewitness reporting, online discussion and networking, and independent investigation projects. Citizen journalism has become an important means for ordinary people to seek information, check facts, set agendas and defend human rights. Thus, it has intervened in and transformed the crisis communication traditionally controlled by the Party-state.

Media and crisis communication in China: Characteristics and turning points

Crisis communication is "the communication between the organization and its publics prior to, during, and after the negative occurrence", and is "designed to minimize damage to the image of the organization" (Fearn-Banks 1996: 2). Effective crisis communication creates a positive reputation for the organization involved and also generates positive public opinion (Sellnow, Seeger, and Ulmer 2002; Sturges 1994). According to Gonzales-Herrero and Pratt (1996), crisis communication should target particular audiences, obtain third-party support, implement internal and external communication, and avoid rumor-mongering. Moreover, it should be a two-way information exchange, rather than a unidirectional information transmission, in order to "[reach] a common understanding of issues" (Shrivastava 1987). Crisis communication is a vital component in managing outbreaks, crises and disasters (Coombs 2007; Coppola 2011; Haddow, Bullock and Coppola 2007). As Marra succinctly puts it, "excellent crisis management cannot exist without excellent crisis communication" (1998: 7). When the social fabric is disrupted by disaster, and people's lives and property are threatened, the government's crisis communication plays an important role in releasing timely and accurate information, reducing social panic, and maintaining social and political stability.

China's crisis communication has long been characterized by lack of transparency, timeliness, and accountability (Liu 1971; Su 1994). Against all principles of optimal crisis communication, such as active communication, transparency and openness, the Chinese government conventionally conceals, minimizes and delays the release of information (Qian 1986; Chen 2008). It was not until November 1979, for example, that China's official media outlet, reported the death of 240,000 people in the Tangshan earthquake three years previously (Xu 2006). Similarly, in the early stage of the SARS epidemic in 2002, the government refused to divulge the exact number of infected people and cracked down on media coverage of the epidemic (Yu 2007b). The assumption behind these routine practices is that crises may cause social chaos. The control of crisis information is therefore perceived as helping to maintain social stability and consolidate the legitimacy of the CCP (Chen 2008).

Mass media, which play an important role in bridging the information gap between the government and the public in times of crisis elsewhere in the world, are heavily controlled in China. They are expected to put the Party's political needs first—that is, to maintain social and political stability. In order to ensure that the media stay on the right track, the State Council and the Publicity Department of the CCP issued the "Announcement on Improving Emergent Incident Reporting" in January 1989, which stipulated that local media outlets were not allowed to report major crises without obtaining approval from senior officials in the State Council. Even when permission was granted, media outlets were required to use the template [*tonggao* 通稿] for reporting disasters that was released by Xinhua News Agency (Chen 2008: 42). In crisis, mainstream media have to work in alignment with the government, propagandize the Party's agendas and guide public opinion [*yindao yulun* 引导舆论]. Jia Li summarizes the function and responsibility that media are conventionally expected to fulfill:

> In crisis reporting, the media should not emphasize a sense of disaster, but a 'sense of the big picture' [*daju yishi* 大局意识] so as to reassure the people, promote social stability, and reduce negative reporting that may induce psychological panic . . . The media should help (the government) rather than make trouble (for the government). (Li 2009: 211)

This instrumentalism is a long-term consequence of the government's highly controlled model of crisis communication. However, the media's role in responding to crises has evolved with the structural transformation of Chinese media as well as the government's social and political reforms. Since 1978, China's authoritarian media system has been gradually transformed into a "propagandist/commercial model" (Zhao 1997). Thus, the Chinese media not only serve the political superstructure as propaganda machinery, but also work as independent entities in the market. Accordingly, in times of crisis the Chinese media must not only serve the Party's need for stability, but also the audience's demand for information.

Sun (2001) points out that since the 1980s, the focus of crisis reporting has gradually shifted from people who participate in rescue and relief work (such as government officials, soldiers and volunteers) to objective crisis information, such as the death toll and victims. Xie, Cao and Wang (2009) conclude that the highly controlled nature of crisis communication has been loosened to a very limited extent by a series of economic and political reforms since 1978. Such changes in crisis reporting practice and crisis management policy between 1978 and the late 1990s did not substantially challenge or change the highly controlled nature of crisis communication. However, they laid the foundation for further crisis communication reform in the twenty-first century.

SARS and China's crisis communication reform

In China, the first real crisis communication reform was triggered by the SARS epidemic in 2003, which transformed the roles of the media, government and public—as well as their interrelations—in times of crisis. SARS started in Guangdong Province, China, in November 2002 and soon spread to Hong Kong and then to Hanoi, Singapore and Toronto. According to the World Health Organization (WHO), by December 31, 2003, the total number of SARS cases had reached 8,096. In China, there had been 5,327 cases, and 349 of those infected had died (*CNN.com* 2004). As a global public health crisis, SARS has been examined by scholars in the fields of crisis communication, crisis management, health communication and public relations. The SARS epidemic marked a turning point in China's crisis communication reform. Three major factors contributed to this reform: pressure from the domestic and international community, investigative reporting by liberal media outlets,[1] and the rise in online communication.

Initially, Guangdong government and the central government dealt with the crisis in the conventional way, covering up the truth and restricting the media's reporting. The Publicity Department of the CCP issued an internal document on February 7, 2003 to central and local media organizations, ordering them to publish uniform opinions and statistics, and to emphasize that SARS had been contained (Xue and Liu 2013). According to Brady (2008: 70), the government's initial mishandling of the SARS crisis was deliberate, and motivated by the need to maintain political stability during the handover between Presidents Jiang Zemin and Hu Jintao.

However, as the situation spiralled out of control, the CCP's leadership came under domestic and international pressure.

In Guangzhou—the capital city of Guangdong Province—and other major cities such as Beijing and Shanghai, rumors about the virus spread swiftly via unofficial channels. The fear of SARS finally developed into widespread panic shopping (Yu 2009: 83). Internationally, the WHO and foreign governments pressured the Chinese government to control the outbreak. Up to 127 countries boycotted Chinese goods and services and temporarily ceased diplomatic, cultural and business activities with China (*Xinhuanet.com* 2003a). The domestic social instability caused by SARS tested the CCP's capacity to govern, and external pressures undermined China's self-claimed status as a "responsible great power". In order to maintain social stability and regain its international reputation, the Chinese government was forced to change its conventional methods of crisis communication.

The liberal media's investigative reporting is another important factor that has driven crisis communication reform. Since the late 1980s, the Chinese media has obtained limited autonomy from marketization and commercialization, particularly local commercial media outlets. They are not as tightly controlled as they were been before, and can occasionally express critical voices. *SMD*, which is based in Guangdong Province, was first to break the government's propaganda directives and reported SARS had not been effectively controlled in March 2003. Following *SMD*, other commercial media outlets in Guangdong and Beijing undertook investigative reporting (Tong 2011), which undid the government's cover-up and informed the public about the real situation, pushing crisis communication reform during the second phase of SARS.

The rise in online communication was also a crucial driver of the reform. When official SARS information was absent, unofficial SARS information circulated rapidly via the Internet and mobile media, such as e-mail, online chat rooms, BBS, and the short message service (SMS) (Yu 2007b; 2009). These mobile and digital media enabled ordinary people to produce, circulate and consume crisis information beyond the mainstream media, creating alternative platforms for spontaneous crisis communication. Such online communication allowed formerly passive audiences to "build their autonomy" (Castells 2007: 249) and minimized the "power gap" (Coombs 1998) between the state and society, thereby challenging the imbalanced power relations institutionalized in China's official crisis communication.

Facing the aforementioned pressures, the government started to adjust its crisis communication practice and attempted to establish an open information system. On April 11 and April 18, 2003, the State Council of the Central Government and the CCP Central Committee issued separate commands that required local governments to report information about SARS cases to the State Council in a thorough and prompt manner (Chen 2008: 41). From April 20 to May 30, the Information Office of the State Council held eight live press conferences to release SARS information and fielded questions from domestic and foreign journalists. From mid-April to mid-May, 2003, the mainstream media's reporting on SARS greatly increased. The *People's Daily*, China's most influential official newspaper, released 143 articles related to SARS from April 18 to April 28—nearly twice as many as the coverage from March 27 to April 17 (Zhu 2005). From May 1 to May 11, CCTV conducted a live broadcast on SARS (Yu 2009). In addition, central and local government bodies began to hold press conferences and release statements via spokespeople (*Sohu.com* 2003; *Xinhuanet.com* 2003b). On May 9, 2003, the State Council released "Regulations for Emergent Public Health Events", ordering local governments to report emergent public health events to the State Council. The government's reformed crisis communication finally led a successful anti-SARS campaign.

Learning from the SARS experience and improving crisis management rose to the top of the Chinese government's agenda. On January 8, 2006, the State Council released a "National Emergency Response Plan" (NERP). The plan categorizes "emergent public events" into four types: natural disasters, industrial accidents, public health and security events. Incidents are categorized on a scale ranging from "ordinary" to "devastating", according to their nature, severity and influence. The plan clarifies the working principles, organizational system and operational mechanisms for dealing with different types of emergent public events, and also requires that the government release complete information in times of crisis (*Xinhuanet.com* 2006). The release of the NERP marked the establishment of China's new system of open information in dealing with various crises. Based on the NERP, the Emergency Response Law was passed on August 30, 2007. The Law offers comprehensive instructions for government institutions on dealing with public crises, marking the institutionalization of crisis management and crisis communication in the everyday work of government.

After a series of reforms, an ideal model of crisis communication has been established at the institutional level. However, to what degree does it work in practice? Has the highly controlled model of crisis communication, with its long historical and political tradition, really been superseded? And what are the characteristics of China's post-SARS crisis communication?

Crisis communication in the post-SARS era: Case study of the 2008 Sichuan earthquake

On the afternoon of May 12, 2008, a 7.9-magnitude earthquake hit Sichuan Province, a mountainous region in western China, killing at least 68,000 people and leaving over 18,000 missing. The Chinese media and government responded in an open, timely and constructive manner.

Xinhua News Agency released the first official report of the quake through its online portal just eighteen minutes after it occurred (He, Li and Wu 2009). Four minutes after the Xinhua news, the CCTV News Channel reported the disaster verbally. From 15: 20, about fifty minutes after the earthquake, CCTV's Channel 1 and CCTV's News Channel interrupted normal programming to commence a live broadcast of a special program, *Focusing on the Wenchuan Earthquake* [*Guanzhu wenchuan dizhen* 关注汶川地震]. The number of dead, injured, missing and homeless all appeared at the bottom of the screen, providing the latest information about the earthquake.

In the following two weeks, the CCTV sent more than 150 journalists to the disaster zones, and covered the rescue and relief work twenty-four hours a day (X. B. Wang 2009). In addition, the State Council held daily press conferences for more than two weeks from May 13, 2008. The government and the state media's proactive crisis communication ensured a transparent flow of appropriate and reliable information. As Wanning Sun commented, the coverage of the Sichuan earthquake was "innovative" and "impressive" (2010: 62). Even foreign media outlets that often criticize China's crisis communication looked at the state media's performance after the earthquake with new eyes. The *Associated Press* described coverage of the earthquake as "a major departure from China's past tendency to conceal crises" (Tran 2008); *The Washington Post* said that China's "normally timid news media" had followed the earthquake with "unprecedented openness and intensity" (as cited in Fan 2008).

From this, it would seem that the government had applied the new model of crisis communication in the Sichuan earthquake. However, Chinese journalists, who work at the front-line in the field and thus have better-informed insights into the rules of China's news, held a different perspective about the government's supposedly proactive response to the disaster. Most believed that the state media's improved coverage of the quake did not prove the Chinese government had totally abandoned conventional information control in times of crisis, but rather showed the government's innovative approach in dealing with crisis in the post-SARS era. For example, Li Datong, one of China's most famous liberal journalists and former editor-in-chief of the *Freezing Point* column in *China Youth Daily*, argued it was mistaken to think that the seemingly transparent earthquake coverage would lead to free crisis reporting in China. In an interview with Spencer (2008) a few weeks after the disaster, Li said:

> During the first week or so, the central propaganda department didn't have time to prepare guidance, so Chinese news media covered the quake by instinct. Now media coverage of the earthquake has returned to its old ways. From now on, we can't expect to get much from reading newspapers or watching television apart from all those heroic deeds. (Spencer 2008)

In an interview with me, a journalist from a provincial TV station who was sent to the disaster area to report the earthquake said that he believed the most important reason for the government taking proactive action after the earthquake was the timing of the 2008 Beijing Olympic Games. He said:

> The earthquake happened less than three months before the Olympics. If the government could not handle the Sichuan earthquake well, China's international image would be damaged. This would affect the upcoming Olympic Games. The 2003 SARS epidemic had already taught the government a vivid lesson. (Personal communication, July 30, 2010)

This informant also mentioned that although the local media were allowed to report the disaster from the scene, the state media (i.e., CCTV and Xinhua News Agency) still dominated crisis reporting. Many news spots were only open for the journalists from the state media. Some important government officials only accepted interviews from the state media. In addition, he complained that many of the news manuscripts he relayed back to the TV station were either censored or rejected by the editor because they were too different from CCTV news. He further argued that:

> Audiences may find that the coverage of the earthquake is more timely and transparent than before. However, they have no way of knowing how the censorship works. As a journalist working inside the journalistic field, I found that the government's control of information and the media's self-censorship are still there. However, the government has started to use a softer and a more strategic way to regulate the media, and talk to the public in times of crisis. (Personal communication, July 30, 2010)

As Li Datong anticipated, the government's control of information soon became much more visible. The initial proactive crisis communication was soon replaced by propaganda directives. Alarmed by the widespread discussion of the number of students who had died in substandard school buildings in the quake zone, the Publicity Department issued orders to rein in local media, and demanded that they focus instead on uplifting tales of heroic rescue and relief work (Spencer 2008). *SMD*, which first disclosed the government's cover-up of the SARS epidemic, was recalled from the disaster area and restrained from covering the schoolhouse scandal. From June 15, 2008, more liberal media outlets were forced to withdraw from the disaster area, in case they investigated the schoolhouse scandal any further (Zhang 2010). The government's cooperative crisis communication had thus given way to conventional media control.

The government's crisis communication in the post-SARS era has shown moderate openness and selective control. The reformed model has been applied, but not fully realized. The conventional mode has been changed, but not abandoned. The proactive and repressive approaches intertwine and characterize China's post-SARS crisis communication. On the one hand, the government has realized that the state media's proactive crisis reporting can help set the tone of media coverage and influence public opinion. The state media, particularly CCTV and Xinhua News Agency, are therefore encouraged and supported to increase their communicative capacities.

This was evident in a speech delivered by President Hu Jintao in April 2008. Having summarized China's experience with disaster reporting over the last five years, Hu said that during any future "public incidents", reporters from CCTV and the Xinhua News Agency must be allowed to report from the scene. He demanded that the Chinese media report crises "at the first available moment" and "increase the transparency of news" (Wang 2008). When celebrating the fiftieth anniversary of the CCTV in December 2008, the CCP's head of propaganda, Li Changchun, urged:

> In reporting important events inside and outside China, we must aim to be timely, open, and transparent. We want to adopt a proactive approach, trying to be the first to get our voice out and communicating our own perspectives. We must work hard to enhance the authoritativeness and impact of our mainstream media. (As cited in Sun 2010: 55)

On the other hand, the government repressed the liberal media's crisis reporting. After the liberal media's victories in the SARS epidemic and the Sun Zhigang case in 2003, political authorities worried that the overwhelming revelation of the negative sides of social reality would impair the image of the nation and pose a threat to the rule of the CCP. Therefore, the liberal media's crisis and investigative reporting have been stymied with various restrictions since 2004. This probably explains why liberal media outlets played a much weaker role in holding authorities to account after the Sichuan earthquake than in the SARS epidemic.

Apart from the innovation of "top-down" crisis communication, the fast rise in "bottom-up" crisis communication is another prominent feature of the post-SARS era. "Bottom-up" crisis communication is not a new phenomenon. In the 2003 SARS epidemic, online rumors and satirical pieces produced by anonymous Internet users imposed effective pressure on the government to open up reporting. However, China only had 68 million Internet users in June 2003 (CNNIC 2003), representing only a limited number who lived in coastal cities. Five years later, the number of Chinese netizens had drastically increased to 253 million (CNNIC 2008), while the number of mobile phone users had also significantly increased from 268.7 million in 2003 to 574.6 million in 2008 (*Askci.com* 2009). The rapid development of the ICTs made it possible for more ordinary people to produce self-generated content, giving rise to the development of citizen journalism. In times of crisis, citizen journalistic practices have formed the main force for intervention in the official crisis communication from the bottom up.

Citizen Journalism as Alternative Crisis Communication in the 2008 Sichuan Earthquake

Scholars have widely studied the practices and functions of citizen journalism in the context of disasters such as 9/11 in 2001, the South Asian tsunami in 2004 and the 2011 Japan earthquake (Allan 2002; Outing 2005; Thurman and Rodgers 2014). In China, because of the government's tight control of news production, the role of citizen journalism as watchdog is more prominent than in the West. Particularly in times of crisis, when the mainstream media's reporting does not satisfy the people's "right to know", citizen journalism tends to fill the vacuum by providing alternative crisis information from below. The 2008 Sichuan earthquake demonstrated the interventional power of citizen journalism in crisis communication. Three forms of citizen journalism were particularly significant in reporting of the disaster, namely eyewitness reporting, online discussion and networking, and independent investigation. Though the three forms of citizen journalism are widely practiced in everyday life in China, in times of crisis, they proliferate and work together to provide alternative crisis communication.

Eyewitness reporting

Chinese citizen journalist relayed the first news of the Sichuan earthquake around the world. Youku (www.youku.com), one of China's main portal sites for audio-visual programs, claimed to have uploaded the first video of the quake at 14: 30, two minutes after it happened. The video was posted online sixteen minutes before the first piece of official news was dispatched by Xinhua (Nip 2009). A blogger from Yunnan Province, which was partially affected by the Sichuan earthquake, wrote of the disaster three minutes earlier than Xinhua news (He, Li and Wu 2009). After the earthquake, eyewitness reports overwhelmed online portals that host citizen journalism, including the Tianya Forum (China's most influential BBS), and Youku and Tudou (China's largest audio-visual websites), as well as blogging sites hosted by major portals (such as www.sina.com, www.sohu.com, www.163.com). The number of blogger videos about the earthquake on sina.com alone increased from 170 on May 13 to 13,170 by 16:40 on May 31, when they had been viewed 6,588,986 times (Nip 2009).

Most eyewitness reporters were residents or volunteers. More than three million volunteers from governmental and non-governmental organizations went to the earthquake zones to participate in rescue and relief work (*Xinhuanet.com* 2009b). These volunteers were mostly urban young people equipped with digital devices. They worked not only as volunteers, but also as citizen reporters, recording what they saw, heard and did.

Zhou Shuguang, a famous blogger, was one of these. He writes under the name "Zola" and has been hailed as "China's first citizen journalist". Zhou became well-known in 2007 for his reporting on "the coolest nail-house" [*zuiniu dingzihu* 最牛钉子户] story. He used a digital camera and blog to record the plight of a married couple's defiance of developers in Chongqing. His blogging soon caught the attention of the traditional media and the nail-house story became a national news event (Xin 2010). Thus established as a citizen journalist, Zola travelled to Sichuan to take part in rescue and relief work three days after the earthquake. He updated his personal blog with eyewitness reporting and his interviews with victims (Zhou Shuguang's blog, May 15–21, 2008).

When crisis events occur, mainstream media seldom have journalists at the scene; it usually takes some time to send journalists and cameramen to the site. In China, mainstream media also need approval from the Publicity Department, or must wait for the Publicity Department to set the tone prior to covering the crisis. This delay frequently causes mainstream media to miss the most newsworthy moments in times of crisis. In contrast, citizens' eyewitness reporting is free from such restrictions and can capture newsworthy moments with portable digital devices. Disseminated online, the self-styled eyewitness reporting has the potential to reach as large an audience as mainstream media. The rise in citizens' eyewitness reporting has challenged the relevance and objectivity of mainstream media in crisis reporting, and has provided alternative information resources to allow audiences to fully understand the situation.

Realizing the advantages of eyewitness reporting, the mainstream media has absorbed this self-styled reporting into their everyday journalistic practice, particularly in crisis reporting. In the aftermath of the tsunami in 2004, BBC News solicited citizen reporting by opening several online columns: eyewitness tales, stories of reunions, photos from survivors, and survivor amateur videos. The UK's *Guardian* compiled some of the best tsunami blogging and published a page of highlights on its website (Outing 2005).

The 2005 London terrorist bombing was the first time citizen reporting entered the mainstream news media. The photos of bomb blasts captured by citizens on their mobile phones, and later published on blogs and photo-sharing sites, appeared in national newspapers and television newscasts around the world the next day (Good 2006). In the aftermath of the Sichuan earthquake, citizen reporting likewise entered China's state media for the first time, as CCTV news programs used a video clip recorded by a student from Sichuan University on his mobile phone and other snapshots which had been widely circulated in QQ chat groups (Nip 2009). The adoption of amateurs' eyewitness reporting in CCTV's crisis reporting enhanced the objectivity and reliability of the "top-down" crisis communication. Though the adoption was sporadic and incidental, it showed the tendency towards the convergence of the citizen and mainstream journalism in China.

Online discussion and networking

Apart from eyewitness reporting, online discussion and networking were also important in relation to citizen journalism in the Sichuan earthquake. Online discussion and networking mean that participants do not need to be physically present at the site of the news event. They discuss disaster-related issues, organize charitable and volunteering activities and become networked via online communication. They are not "the people formerly known as audience" (Rosen 2006), who passively wait to be informed by the media. Instead, they actively seek, circulate, compare, check and discuss crisis information on the Internet, generating networked citizen journalism.

By observing the Tianya Forum in the aftermath of the earthquake, Nip (2009) found that people raised issues of concern which were not addressed by mainstream media, and discussed them with other Internet users. Most issues were critical and beyond the agendas of the mainstream media, for example, whether the prediction of the earthquake was covered up, and whether the collapsed school buildings were "tofu-dreg" projects (that is, constructed so poorly that they were as structurally weak as tofu). Reese and Dai (2009) also discovered that some citizen postings in online forums criticized, supplemented, commented, checked or challenged the accountability of mainstream journalism, evoking heated discussion on the objectivity and authority of official crisis reporting. Moreover, many citizen postings aimed at organizing, mobilizing and networking non-governmental rescue and relief work, such as online donations, mourning and volunteer recruitment (Qu, Wei and Wang 2009). Online discussion and networking had become the most important means for the majority—who were not at the scene—to participate in crisis communication. It reflected public concerns, needs and opinions, and helped to draw non-government forces in to the post-disaster rescue and relief efforts.

Eyewitness reporting and online discussion and networking represent citizen journalism. Both of these modes are usually spontaneous, and can be executed by ordinary people with digital devices and Internet access. However, a third form of citizen journalism, independent investigation, is more planned, organized, elite-based and time-consuming. It investigates sensitive issues that the government intentionally covers up in crisis. The investigation is often a process of championing the rights of marginalized social groups and is therefore more risky and transformative than other forms of citizen journalism.

Independent investigation

As previously mentioned, school construction was a sensitive topic in reporting of the Sichuan earthquake. In the earthquake, over 7,000 schoolrooms collapsed and killed approximately 5,335 students. The large number of student casualties led to widespread discussion about the quality of the school buildings. This, in turn, led to allegations of corruption against government officials and contractors, who were said to be complicit in the construction of substandard school buildings and to have pocketed the resultant surplus funds. The school construction scandal soon became a focal point of earthquake reporting. Contrary to its initial openness, the government downplayed the scandal and suppressed the media's investigations. When investigative journalism was restricted, social activists, public intellectuals, and activist bloggers took up the cause as independent citizen journalists. They used the Internet and other digital media to expose, investigate and report the scandal from independent perspectives, providing the most radical crisis communication related to the Sichuan earthquake.

Of these independent investigations, the civil rights advocate Tan Zuoren's had the greatest social influence in China and overseas. After the earthquake, he came up with a proposal called the "5.12 Student Archive" [*Wuyier xuesheng dang'an* 5.12 学生档案], which aimed to set up a database of student victims. Tan travelled to the disaster areas to collect the relevant data, and called for more volunteers to participate in the investigation.

On March 28, 2009, just as Tan was ready to release his database, he was detained. He was formally accused of defaming the CCP in email exchanges with overseas dissidents regarding the 1989 Tiananmen student movement. On February 9, 2010, he was sentenced to five years in prison for "inciting subversion of state power" (Wang 2010). As a result, Tan Zuoren's investigation could not be released publicly. However, his investigation inspired two subsequent independent investigations: Ai Xiaoming's citizen documentary project and Ai Weiwei's citizen investigation project. These two projects released their investigation findings and demonstrated the power of independent investigation in times of crisis.

Ai Xiaoming's citizen documentary project

Ai Xiaoming, born in December 1953, is a professor at Sun Yatsen University in Guangzhou. She is a documentary filmmaker and human rights activist dedicated to making documentaries promoting the rights of women and HIV victims, and films on other controversial social issues in China. In the aftermath of the Sichuan earthquake, Ai Xiaoming focused her camera on the furore surrounding school construction. In an interview with *Asia Weekly*, Ai said that several weeks after the earthquake, one of her friends, who worked as a volunteer in Sichuan, gave her a call and told her that many media outlets had been forced to withdraw from the quake zones since June 15, 2008 in case they reported the school construction scandal (Zhang 2010). Ai's long-term interest in China's women and children motivated her to go to the disaster area in search of the truth.

In June and August 2008, Ai Xiaoming visited most of the counties that had been severely affected by the quake. She reached the ruins of collapsed school buildings and interviewed parents of student victims while subject to various restrictions from the local government. She recorded what she saw and heard by DV, and also collected rich visual materials filmed by local people. In early 2009, Ai finished her first documentary about the school construction tragedy, *Our Children* [*Women de wawa* 我们的娃娃]. In the 73-minute documentary, she interviewed the parents of student victims, independent scholars, Internet writers, geologists, environmentalists and human rights lawyers, trying to draw as many unofficial voices as possible to speak about the quality of the school buildings (Ai, X., May 12, 2010).

Ai's documentary can be considered to be part of China's "New Documentary Movement" [*Xin jilupian yundong* 新纪录片运动] (Lü 2003b), also known as the "Independent Documentary Movement" [*Duli jilupian yundong* 独立纪录片运动] (Berry 2007). The movement refers to a wave of artistic film practice initiated by professional or semi-professional filmmakers in the early 1990s which represents China's marginalized social groups and advocates social justice and equality through documentation (Berry 2007; Lü 2003b). Nevertheless, Ai Xiaoming prefers to define her documentaries as journalistic practice. In an interview with Viviani (2010), Ai said she is not interested in film style or technique, but whether her work effectively conveys newsworthy information to her audiences. As she once argued in an academic dialogue:

> The problem of the Chinese media is their insufficient power in supervising the government. Many social groups are underrepresented. We need to develop various theoretical supports to solve this problem, including how to use media, expand media, and develop citizen journalism. (As cited in Ma and Zhou 2009: 207)

Ai's documentation constitutes her effort to solve the problems in the Chinese media. From her point of view, documentary cinema is an example of citizen media. Making documentaries *per se* is a strategy to represent oppressed voices. Therefore her documentaries bear the hallmarks of investigative journalism, such as the watchdog role, agenda-setting function and mobilization. This can be seen from her aspirations for *Our Children*:

> I hope this documentary is of social service. After seeing the documentary, the audiences' original understanding about the Sichuan earthquake might be challenged. They will discuss the issues my documentary represents, and ask why these issues exist. During the

> process of discussion, some invisible issues could become visible . . . *Our Children* does not answer the question as to whether the collapsed schoolhouses are tofu-dreg projects in a direct way. It leaves the question to the audience. It encourages the audience to consider: why is it difficult to ascertain whether the schoolhouses are substandard or not? Do we need to ascertain this? If so, what is the greatest force that prevents us from approaching the truth? (As cited in Zhang 2010)

Ai's documentary was a collaborative project. *Our Children* contains not only Ai's DV recordings from the scene, but also incorporates external visual materials, such as digital photos shot by the victims' parents and anonymous citizen journalists. Ai Xiaoming's role in documentary production resembled that of an investigative journalist. However, her non-journalist identity enabled her to investigate the issue more freely than investigative journalists, who were imposed with institutional restrictions. But at the same time, the lack of support and protection from an institutional media entity made her more vulnerable than investigative journalists if she provoked the ire of authorities.

Our Children could not be released officially or displayed publicly, in part because its production coincided with the increased political sensitivity surrounding the twentieth anniversary of the 1989 Tiananmen student movement. Ai Xiaoming was blacklisted for her investigation into this forbidden issue. Her personal blog, based on sina.com, was blocked on October 19, 2009; also in mid-October, Ai was prevented from going to Hong Kong to attend the Seventh Social Movement Film Festival with *Our Children* (*Radio Free Asia* 2009).

Lacking public and legal channels to disseminate her documentary, Ai Xiaoming (September 26, 2010) found three alternative ways to disseminate her works. First, the volunteers and interviewees who participated in the production were given free copies, and were encouraged to reproduce and distribute them widely. The second channel was through libraries at home and abroad. Collected by libraries' visual archives, her documentaries could be seen by academics, university students, journalists and NGO workers. This earned her royalties, which could support her future works or the projects of other independent documentary makers. The third and the most important channel was the Internet. Ai and some other proactive audiences who had copies of the documentary uploaded it for free viewing, circulation and download. Although the uploaded videos and shared links were repeatedly removed or blocked from domestic cyberspaces, they could flow into foreign websites beyond the control of the Chinese government, such as YouTube, Twitter and Facebook. The transnational circulation of her documentary made the domestically–suppressed scandal visible to international audiences, media organizations and human rights advocacy groups, turning it into a global news event.

Ai Xiaoming's post-earthquake documentaries have developed what Chris Berry (2003) calls the "socially engaged" mode of independent documentary making by incorporating strong journalistic features. Ai herself (2008) sees them "not as art but as propaganda and as agent for change". In this sense, *Our Children* is not merely a documentary film, but an instance of investigative citizen journalism presented in documentary form. It has expanded the information network to bridge the information gap and start conversations between disadvantaged social groups, mainstream media, civil society and government, engaging both state and non-state players in the otherwise downplayed social issue.

Ai Weiwei's citizen investigation project

Like Ai Xiaoming, Ai Weiwei investigated the school construction scandal following the Sichuan earthquake, but recorded his findings using a different new media technology—the weblog. Ai Weiwei, born in May 1957, is a contemporary Chinese artist and intellectual activist with an international reputation. He is the son of one of China's most famous poet, Ai Qing and contributed to the design of the Birds' Nest Olympic Stadium. He has been highly and openly critical of China's record on human rights and democracy. After the earthquake, Ai Weiwei was unsatisfied with the government's refusal to release the student death toll from the earthquake, and so resolved to compile a list of children killed in the school buildings by means of independent investigation.

In an interview, Ai Weiwei criticized the government's deliberate silence concerning the student victims and the school construction scandal. He argued that "all citizens should have the right to supervise the government, as well as the responsibility to investigate the truth when the government keeps silent". The main purpose of his project was to "show respect to every individual victim's life and refuse to forget the tragedy" (Wu 2009).

On December 15, 2008, Ai Weiwei's Citizen Investigation Group (CIG) was founded in his studio in Beijing.[2] The CIG started collecting profiles of student casualties through limited online information, such as memorial websites and reports from NGOs. In order to obtain complete information, CIG sent four volunteers to Sichuan to collect students' profiles on January 17, 2009. They visited twenty-one villages and towns, interviewed the parents of the student victims, and shot scenes of the rubble. They emailed the fruits of their labor to the head office in Beijing. Based on this information, Ai Weiwei produced his first investigative report, which was released via his personal blogs on Sina and Sohu on March 15. Tianya Forum soon reposted the report, which gave rise to heated online discussion. Thus, the suppressed schoolhouse construction scandal re-entered the public discursive sphere through alternative online communication. Ai Weiwei's blogs soon became a source of reliable information for those who wanted to know the truth of the issue.

Having attracted significant attention, Ai Weiwei began to use blogs to mobilize the public to participate in his project. On March 20, he posted an advertisement to recruit volunteers. The ad received considerable feedback from the audience, and from March 25 to April 21, three batches of volunteers (eighteen, nine and eleven people, respectively) were recruited and sent to Sichuan. Ai Weiwei kept blogging the first-hand information sent from the front-line. From March 21 to May 29, Ai Weiwei posted 202 blog entries that named student victims and 115 entries that documented the volunteers' investigation. His blogs accumulated over ten million viewings. In addition, his blog entries were widely circulated and reposted on BBS, blogs and instant messaging platforms, such as QQ and MSN, making his investigation a hot Internet event.

In light of the political sensitivities within China, Ai Weiwei enlisted overseas media outlets to publicize the results of his investigation. By May 8, 2009, he had accepted nearly seventy interviews from media such as NBC, BBC, Reuters and NHK, which put the Chinese government under pressure. Under attack both internally (from cyberspace) and externally (from the international community), the government finally released the death toll of the schoolchildren on May 7, 2009. A total of 5,335 students were finally confirmed dead or missing in the Sichuan earthquake (Ang 2009). However, the names of the lost schoolchildren were not attached with the official figure.

Due to the great social influence of Ai Weiwei's investigation and blogs, the government shut them down on May 29. Having lost the blogging platforms in domestic cyberspace, he moved his blog overseas. He reposted all blog entries on *Bullogger.com* [*Niubo wang* 牛博网], a Chinese political website with a server based in the US, and continued blogging.[3] Though the website was also blocked in China, skilful Chinese netizens, who knew how to circumvent the Great Firewall, could access the site and repost his blog entries in China's cyberspace, enabling the reverse flow of the blocked information. Millions of netizens reposted Ai Weiwei's blog entries to support his citizen investigation. By the end of July 2009, the investigation was largely completed. A total number of 5,194 students were confirmed missing or dead—close to the official figure. Moreover, a list of 4,851 students was compiled and released on his blog.

In Ai Weiwei's citizen investigation project, blogs were the central platform for releasing the investigation report, interacting with audiences both online and offline, mobilizing public opinion, and building national and transnational advocacy networks. Ai's strategic use of blogging reflects the main characteristics of global networked social activism. As Jeffrey Juris points out, in networked social movements, "activists have used new digital technologies to coordinate actions, build networks, practice media activism, and physically manifest their emerging political ideas" (2005: 192). In Ai Weiwei's project, blogging, albeit under censorship, functioned similarly to other forms of engagement via new media in global social movements, which are both "symbolic and material" (Lievrouw 2011: 158). It represented suppressed agenda as alternative crisis communication, and also defended the rights of marginalized groups in a relatively safe mode of interaction and engagement centered on the Internet.

The characteristics of independent investigation

As Ai Xiaoming's and Ai Weiwei's investigations reflect, independent investigation has a higher threshold of expertise than eyewitness reporting and online discussion and networking. The investigators are usually experienced social activists and public intellectuals rather than ordinary people. Their professional knowledge and rich experience of organizing rights defense actions ensure that they discover the topics with the greatest public interest for investigation in crisis. Their established reputation and social influence increase the credibility of their investigation and help gain support from national and international communities.

Ai Xiaoming and Ai Weiwei represent the new public intellectuals in China's Internet era. In contrast to public intellectuals in the 1980s and 1990s, who spoke only in their specialized knowledge communities, they speak for marginalized social groups and aim to "[translate] their community's language into a public language" (as cited in Cheek 2006: 414). Their investigations resonate with Xu Jilin's expectations of public intellectuals in China's current fragmented and pluralistic society. As a leading historian of modern China, Xu encourages them to "create a public by devising a language of translations—a discourse", which "reflects, acknowledges and engages in this diversity" (as cited in Cheek 2006: 414). Ai Xiaoming and Ai Weiwei used documenting and blogging as a means to translate the languages of vulnerable social communities into a public discourse. These mediated forms of advocating social change echo Douglas Kellner's suggestion directed to new public intellectuals in the West:

> To be an intellectual today involves use of the most advanced forces of production to develop and circulate ideas, to do research and involve oneself in political debate and discussion, and to intervene in the new public spheres produced by broadcasting and computing technologies. New public intellectuals should attempt to develop strategies that will use these technologies to attack domination and to promote education, democracy and political struggle —or whatever goals are normatively posited as desirable to attain. (Kellner 1997: 24)

However, these new public intellectuals are usually in precarious situations in China. As Ai Xiaoming has said, no matter who you are in China, "as long as you stand on the side of the marginalized groups, your own social status will quickly become marginalized" (Ma and Zhou 2010). Ai Xiaoming was denied passport renewal and restricted from going abroad (Hutzler 2010), and Ai Weiwei was detained by police in Beijing for more than two-and-a-half months from April 2011(Branigan 2011). Such unfair treatment was due to their long-term rights defense actions for marginalized social groups and open criticism of the CCP.

Independent investigation requires not only the wisdom and courage of the new public intellectuals, but also public participation and collaboration. *Our Children* included many interviews with parents of student victims and eyewitness recordings from anonymous citizen journalists. Ai Weiwei's investigation could not have been finished without the support of more than forty volunteers. The impact of both projects rested on numerous netizens sharing, reposting and commenting on them. Thus, in China today, the concept of citizens' rights is no longer institutionalized and territorialized, discussed only among public intellectuals and social activists. Rather, defending these rights has become an open, collective and participatory practice, through which people of different social classes and statuses advocate social justice and changes via networked communication.

Moreover, independent investigation is a long-term task which may take years. This has challenged the notion that crisis communication is finished immediately after crisis events. In China, where official crisis communication usually has blind spots, addressing covered-up issues through unofficial channels takes time and thus has extended the timeframe of crisis communication. After finishing *Our Children*, Ai Xiaoming made another three documentaries related to the scandal. *Citizen Investigation* [*Gongmin diaocha* 公民调查, completed December 2009] and *Why Are the Flowers So Red?* [*Huaer weishenme zheyang hong* 花儿为什么这样红?, completed April 2010] focus on Tan Zuoren's investigation and Ai Weiwei's citizen investigation respectively. In September 2010, Ai Xiaoming finished her fourth documentary about the Sichuan earthquake, *Forgetting Sichuan* [*Wangchuan* 忘川], in which she records memorial activities on the first anniversary of the Sichuan earthquake, and tells how the parents of student victims have gradually accepted the reality and started new lives.

Ai Weiwei similarly started using documentary and other artistic forms to follow up on his investigation of the school construction scandal. In September 2009 he released the documentary *Disturbing the Peace* [*Lao ma ti hua* 老妈蹄花], which related the difficulties he encountered when he attempted to testify during the trial of Tan Zuoren the previous month. In 2010, Ai Weiwei released two artistic audio-visual works, *4851* and *Remembrance* [*Nian* 念]. The former is simply a long written list of student casualties in the Sichuan earthquake accompanied by music. *Remembrance* is a sound work co-produced by thousands of volunteers, with each reading out the name of a student victim.

Ai Xiaoming and Ai Weiwei's projects intersect. Ai Weiwei's citizen investigation project is the main subject of Ai Xiaoming's documentary *Why Are the Flowers So Red*? Ai Weiwei's *Disturbing the Peace* could be read as a sequel to Ai Xiaoming's *Citizen Investigation*—both focus on the Tan Zuoren case. These projects work together to investigate the school construction scandal from different perspectives, thereby demonstrating the collaborative power of media activism, and facilitating a cross-textual reading of rights defense actions in the aftermath of the Sichuan earthquake.

Summary

The characteristics of China's post-SARS crisis communication can be deduced from responses to the 2008 Sichuan earthquake. First, official crisis communication has been innovated in order to adapt to China's changing social, political and technological environments. The government and state media now respond promptly and proactively in times of crisis. However, the increased openness of the information system has not yet displaced the conventional information control; minefields still exist in crisis communication. Second, the rise of citizen journalism has enabled various social groups to participate in crisis communication from the "bottom up". This alternative crisis communication has redressed the information gap and challenged the asymmetrical power relations institutionalized in crisis communication. The contestation, negotiation and interaction between the"top-down" and "bottom-up" crisis communications has been the most prominent feature of China's post-SARS crisis communication.

Due to the rise of alternative crisis communication in the form of citizen journalism, crisis events in China are no longer unilaterally selected for coverage by national broadcasters and scripted in accordance with the Party's propaganda directives. Rather, they have become what Fisk (1994) calls "discursive events" and Lévi-Strauss (1966) calls "hot moments", which see different discursive communities drawing on different symbolic resources to interpret certain moments, and compete for the discursive power to represent them. Crisis communication has become a contested field in which the CCP aims to legitimize institutional power and symbolize social cohesion through innovated communicative practices, while citizens fight for human rights and political change through digitized activism.

4

Media Scandal: Online *Weiguan* as Networked Collective Action

The Chinese term "*weiguan*" [围 观], literally translated as "surrounding gaze" in English, refers to a crowd activity that occurs in public venues. It describes mass gatherings around public spectacles, such as ritual celebrations, traffic accidents and public executions. In this sense, *weiguan* is a common human behavior that exists in all cultures and societies. However, with the penetration of the Internet in Chinese people's everyday life, the term has been used as a cultural metaphor to vividly describe a type of mediated mass mobilization in China's cyberspace. Now, Chinese netizens and the media use the term "online *weiguan*" [*wangluo weiguan* 网络围观] to refer to the online outpouring of public opinion around officials' wrongdoings and power abuses. This public display of emotion and opinion tends to form networked power and group pressure to promote transparency surrounding scandalous events, and their resolution.

This chapter discusses how online *weiguan* has intervened in and transformed both official and unofficial investigations of political scandals in China. The chapter is structured in two main sections. The first section discusses the rise and fall of investigative journalism in China's post-reform period. The second section examines how online *weiguan*—a networked collective action beyond state-controlled investigative journalism—discloses, disseminates and discusses political scandals and sets items on the government's agenda. In brief, online *weiguan* has become a mediated form of political participation through which ordinary Chinese people can fight power abuses by officials. This virtual action has demonstrated the collective intelligence and crowd politics of Chinese netizens, and provided impetus for political reform.

Media and Scandal: Revisiting Investigative Journalism in China

Scandal is ubiquitous in human society. In terms of morality, they reveal a moral order that is temporarily disrupted (Jacobsson and Lofmarck 2008: 205). From a political point of view, scandals are abuses of power "at the expenses of process and procedure" (Markovits and Silverstein 1988: 6–7). On the basis of the different subjects, processes and public responses they involve, scandals are classified into three types: official corruption, human rights violations and celebrity scandals (Waisbord 2004). In media-saturated societies, scandals have become more "media-centric". As Thompson argues, "the media become the principal mechanism through which corruption is made visible to others" (1997: 51). The media function not only as storyteller, but also as director, which controls the drama and generates various media scandals.

According to Lull and Hinerman, media scandal occurs "when private acts that disgrace or offend the idealized, dominant morality of a social community are made public and narrativized by the media" (1997:3). The term underlines the media's influence on the emergence, development, prominence and consequences of scandalous events. Media scandals are therefore "in varying ways and to some extent, constituted by mediated forms of communication" (Thompson 1997: 49). Among the various mediated forms of communication, investigative journalism is most important in exposing, reporting and investigating scandalous events.

In the Western tradition of media, investigative journalism is regarded as the "most vigorous" journalistic practice (Glasser and Ettema 1989). This tradition emerged with the rise of scandal politics in the 1960s and 1970s in Western democracies. The "Watergate Scandal" in the US and the "Thalidomide Scandal" in the UK are representative cases of successful investigative journalism during that period. Investigative journalism embodies the professional ideology of journalists as "kings without crowns", as well as the power of news media as "the fourth estate". It is often regarded by journalists and media commentators as the paradigm of "good journalism" (Pilger 2005), and is vital to "the checks and balances of a healthy democracy" (Ricketson 2001).

However, investigative journalism in China has a relatively recent history. As a genre of modern journalism, it emerged in China after the implementation of Deng Xiaoping's economic reforms in 1978

(Chan 2010). Its practice is influenced by various philosophical principles rooted in China's ideological traditions, such as Confucianism, liberalism and Communist Maoism (Tong 2011: 15). It differs from Party journalism or "command journalism", because it challenges the role of the Chinese media as the CCP's propaganda apparatus. It has created a public channel for Chinese media to represent public interests, while monitoring the CCP's governance and policies. It advocates and aims to perform "supervision by public opinion" [*yulun jiandu* 舆论监督].

Supervision by public opinion, according to Cheung (2007: 1), refers to a dynamic and interactive process involving the CCP, the media and the public. It emphasizes the media's positive function in mobilizing public opinion to act as a check against the state, defining substantive social problems, and pushing for legislative or policy reforms. Investigative journalism plays a central role in realizing the media's function as a watchdog of China's economic and political reforms. This journalism evolves with China's changing political atmosphere, deepening media reform and rapid development of media technologies.

The rise of investigative journalism was one of the most significant developments in Chinese journalism during the 1990s (de Burgh 2003; Tong 2011; Zhao 2000). There were two main reasons for this rise. One was the changing media landscape in China. Since the late 1980s, the government had pushed the media to become commercially profitable. It was difficult for media to survive by doing Party journalism alone under increasingly fierce market competition; a new genre of journalism was needed to attract both advertisers and subscribers. In Western societies, the commercialization and tabloidization of the press led to the production of eye-catching scandalous news to increase the circulation of newspapers. As Thompson (1997: 49) succinctly put it, "scandal sells". Similarly, the rise of investigative journalism in the 1990s in China was partially due to the market value of scandalous news.

The other reason for the development of investigative journalism was the political need to "reassert control over an unruly and dysfunctional bureaucracy" created by economic reform (Zhao 2000: 580). Since 1978, the prioritizing of economic growth above all other issues had caused many social problems; to name a few, massive lay-offs in state-owned enterprises, the geographical displacement of farmers, the unfettered abuse of taxation powers by local government, and the drastic increase in local corruption.

These controversial social issues threatened the legitimacy of the reforms and the stability of the CCP's governance. Investigative journalism was therefore welcomed by the CCP, because it could monitor local governments, promote public trust in leadership, and create a positive environment for further reform. Against this backdrop, investigative journalism quickly became a popular news genre in both central and local media organizations in China from the 1990s.

Investigative journalism in China was commissioned by two main organizations: CCTV, and the newspapers owned by the Guangzhou-based Southern Daily Group (Tong and Sparks 2009). In 1994, CCTV launched its daily investigative program *Focus Interview* [*Jiaodian fangtan* 焦点访谈], which featured hard-hitting reports on corruption and government wrongdoings. Newspapers of the Southern Daily Group were the first among China's print media to embark on investigative reporting. In 1995, *Southern Weekend* began to engage in in-depth critical reporting. *Focus Interview* and the investigative reporting of *Southern Weekend* proved highly successful in the marketplace. The former quickly became one of China's most popular TV programs, attracting approximately 300 million viewers every day (Chan 2010: 8). The latter soon became one of the most influential newspapers in China and established its brand among readers because of its independent and sharp style. Their success inspired a nation-wide tide of investigative programs, columns and media. Among them, *Freezing Point* column in *China Youth Daily* (1995), CCTV's *News Probe* (1996), *Southern Metropolitan Daily* (1997), Shanghai TV's *Journalists' Investigation* (1998), and *Caijing* magazine (1998) gained a national reputation for investigating official corruption, wrongdoings and cover-ups.

However, the heavily-controlled nature of China's media system means that the supervisory function of Chinese investigative journalism does not correspond to the press theory of the "fourth estate" in the West, which implies that the media has independent power to criticize the state. Rather, investigative journalism in the PRC can only focus on "specific issues", deal with "concrete problems", and is therefore "practical-minded" (Zhao 2000: 579). Metaphorically speaking, it hunts down "flies, not tigers".[1] It exercises supervisory power only at the behest of the Party and works as the watchdog "on the Party's leashes" (Zhao 2000). Therefore, its power is quite "precarious" (Tong and Sparks 2009: 337), due to China's changing political and economic environment.

After a short period of vigorous development, investigative journalism in China has declined since 2003. Tong (2011: 31) attributes the decline to the same reasons as its rise, that is, the Party's political needs and the further marketization of China's media industry. In 2003, Hu Jintao and Wen Jiabao came to power and initiated what the Chinese media dubbed "Hu-Wen New Deal" [*Huwen xinzheng* 胡温新政]. The Hu-Wen leadership aimed to construct a "harmonious society" [*hexie shehui* 和谐社会] by mitigating domestic social conflicts. Thus the media's role as a propaganda machine for the Party and an educative tool for ideological unity was reasserted.

This was especially the case after the SARS epidemic and the case of Sun Zhigang in 2003, in which investigative reporting by *SMD* and *Southern Weekend* forced the government to reform its crisis communication and abolish the detention and repatriation system. Following these events, the CCP stepped up its control of news media and suppressed investigative reporting, in case the media's growing power in public opinion supervision would challenge the CCP's legitimacy. Hence "legal, administrative and extralegal means" were used to crack down on investigative journalism (Chan 2010: 10). For example, in 2004, the Southern Daily Group, which owned the *SMD* and *Southern Weekend*, suffered a reprisal: the editor and deputy-editor of the *SMD* were arrested for alleged corruption. In the same year, the CCP's Publicity Department banned the media from conducting "extra-regional media supervision" [*yidi jiandu* 异地监督].[2] In January 2006, *Freezing Point*, the "critical idea" column in *China Youth Daily*—the official newspaper of the Communist Youth League, was temporarily shut down for "rectification" by the Publicity Department of the Communist Youth League due to the publication of a critical feature that criticized China's official history textbooks for teaching an incomplete history of China's Qing Dynasty, that fostered blind anti-foreign sentiment and nationalism (Pan 2006).

The decline of investigative journalism was also due to "new kinds of censorship" caused by the "maturing media market" (Tong and Sparks 2009: 338). The high cost and political risk of engaging in investigative reporting forced many established media outlets to reduce their budgets and start to explore new, profitable journalistic practices in order to survive in the increasingly competitive media market. However, the political and economic pressure did not stop investigative journalism or make it disappear. Instead, it has been evolving and adapting to new social realities, particularly the development of new media technologies. The popularity of networked online communication promises to reenergize Chinese investigative journalism (Chan 2010). This new form of communication has provided an alternative means to expose, report, disseminate and discuss official wrongdoings beyond the traditional television and press, changing the traditional mobilization model of investigative reporting.

Previously, official wrongdoings were first identified by mainstream media. Their reporting was likely to elicit responses from the public and shape public opinion, which in turn forced the government to resolve the exposed problems. However, when the mainstream media's investigative reporting was restricted, corruption was brought to the public's attention through online tip-offs. Once such matters amassed popular interest online, they were likely to catch the attention of mainstream media and prompt an investigation. This follow-up reporting could reinforce the existing online public opinion and push the government to address the scandal. The "contra-flow of agendas" (Tong 2011) or "reversed agenda-setting effects" (Jiang 2014), enabled by networked online communication, have transformed the mobilization model of investigative reporting. Figures 4.1 and 4.2 modify Tong's (2011: 200) mobilization model of investigative reporting in China:

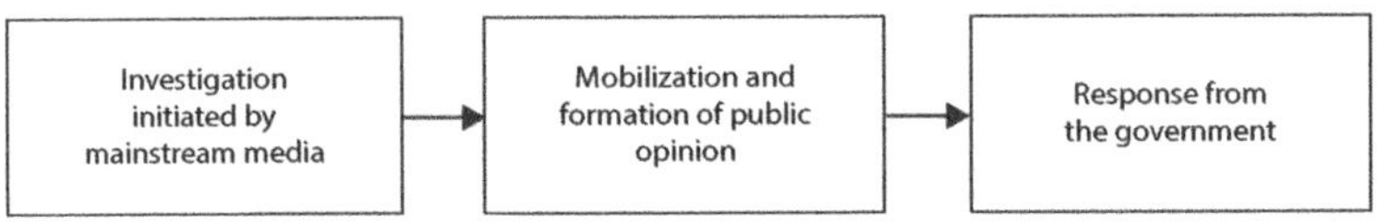

Figure 4.1 Mobilization model of traditional investigative journalism

Figure 4.2 Mobilization model of Internet-mobilized investigative journalism

Figures 4.1 and 4.2 indicate that the mobilization and formation of public opinion, either offline or online, is a determining factor in advancing agendas to a higher level. As the political slogan "supervision by public opinion" reflects, public opinion is the subject that exercises the supervisory power. Particularly in Internet-mobilized investigative journalism, public opinion support is critical. The stronger online public opinion is, the more attention an issue will gain from mainstream media. Popular support for investigative reporting on an issue also makes it less politically hazardous, and means that the government is more likely to resolve the scandal.

Online *Weiguan*: Mediated Mass Mobilization in China's Cyberspace

Online weiguan*: Historical origins and characteristics*

The networked collective action of *weiguan* in the cyberspace greatly contributes to the mobilization and formation of online public opinion around offline political scandals. In fact, *weiguan* is not a new invention of the Internet age, but has its origins in China's cultural and political history. China's literary giant Lu Xun probably first wrote about the *weiguan* phenomenon in his short novel *Medicine* [*Yao* 药] in 1919. The story is about Old Chuan buying a steamed bun dipped in blood at an execution site in order to save his son's life. Old Chuan held the superstitious belief that steamed buns dipped in human blood could cure his son's tuberculosis. In this story, Lu Xun vividly described a crowd of indifferent Chinese who dumbly watched the beheading of the revolutionist Xia Yu, who had fought for their liberation and freedom:

> Craning their necks as far as they would go, they looked like so many ducks held and lifted by some invisible hand. For a moment all was still; then a sound was heard, and a stir swept through the on-lookers. There was a rumble as they pushed back, sweeping past Old Chuan and nearly knocking him down. (Lu [1919] 1960: 26)

Lu Xun used Xia Yu to allude to Qiu Jin [秋瑾], a pioneer of the 1911 revolution who advocated the overthrow of the Qing Dynasty and the establishment of a republic. By telling this story, Lu Xun criticized the "culture of gaze" [*kanke wenhua* 看客文化] in Chinese society. To Lu Xun, the culture of gaze, exemplified in the *weiguan* crowd at Xia Yu's beheading, was one of the Chinese nation's deep-rooted weaknesses. It reflected the ignorance, numbness, conservativeness, and backwardness of ordinary Chinese in pre-modern society. They did not care about the fate of the nation and were ignorant about revolution. They were obedient subjects, though they suffered a lot under imperial rule. They did not want to take part in the revolution, because they were afraid that rising up against authority would be risky and disrupt their routine, family-centered lives. However, they were curious to see the result of the revolution as spectators, and gossiped about the revolution in private. Lu Xun also criticized the 1911 revolution for its lack of grounding in popular support, and its ultimate failure to deliver substantial improvements to people's lives. He thought that if only the revolutionary movement could unite ordinary Chinese and turn them from passive onlookers into active participants, it would succeed.

Lu Xun's prediction came true in the Chinese Communist revolution led by Mao Zedong. In political campaigns from the 1930s to the 1970s, passive *weiguan* spectators were strategically organized and mobilized by the CCP and transformed into masses with revolutionary fervor. A series of techniques were consciously used by the CCP to harness the emotional energy of the *weiguan* masses in order to realize its political purposes in such practices as "speaking bitterness" [*suku* 诉苦], "denunciation" [*kongsu* 控诉] and "criticism-self criticism" [*piping yu ziwo piping* 批评与自我批评] (Perry 2002). These techniques were intentionally applied in "show-and-shame" *weiguan* contexts in order to stir up outrage against the CCP's political targets and heighten people's emotional commitment to the CCP's mass movements.

The mass criticism meetings held during the Cultural Revolution were a case in point. In these meetings, Red Guards, Rebels and others became *weiguan* crowds. They stood around "counter-revolutionaries" to hold open trials of targets selected by the CCP. In stark contrast to the numb and indifferent onlookers at Xia Yu's execution in Lu Xun's novel, these crowds became frenzied participants infected with revolutionary fervor. They shouted slogans, denounced crimes or even threw stones to attack their targets. Their emotions were intensified by the strategic work of the CCP. The "show-and-shame" ritual, as exemplified in the mass criticism meetings of the Cultural Revolution, proved to be an effective strategy for the CCP to propagate and mobilize mass emotion, and punish political enemies at different stages in the Mao era.

This ritual continued to exert powerful influence in the CCP's political governance in the post-revolutionary era. After the Cultural Revolution, the CCP shifted its focus from the Communist revolution and class struggle to the maintenance of social and political stability, and the pursuit of economic development. The "show-and-shame" ritual was used again to punish destructive forces that harmed social and political stability. In "campaigns against criminal activities" [*yanda* 严打] since the 1980s, the "show-and-shame" ceremonies have taken the form of public trials and shame parades. In these *weiguan* contexts, the masses are not encouraged to participate as mobs in the Communist revolution, but are organized as voluntary spectators of public executions. By openly showing the punishment of criminals, the CCP aims to display its sovereign and disciplinary power as China's legitimate ruling Party, while warning the populace to behave within the confines of the law. For example, on 29 November 2006, the Public Security Bureau in Shenzhen, Guangdong Province, handcuffed approximately one hundred prostitutes and forced them to march along a street in front of a jeering crowd. The march was broadcast via television to show the government's efforts to combat the illegal sex trade (Watts 2006). The visibility of the punishment of the few was used to show the disciplinary power of authority over the many (Foucault 1977).

The "show-and-shame" ritual and the *weiguan* context it requires have been consciously and strategically used by the CCP to achieve its political purposes over the past eighty-odd years. In recent times, however, this ritual has been reinvented and reversed by ordinary Chinese to articulate social critiques and supervise governmental power, thereby forming a popular online *weiguan* phenomenon. Accordingly, the term of *weiguan* has been redefined and has become a synonym for active participation. As communication scholar Hu Yong argues, online *weiguan* is a kind of "minimal (or bottom-line) form of public participation" in Chinese society, in which free expression and political participation are restricted (as cited in the China Media Project 2011). It has become an alternative way for ordinary Chinese to exercise supervision by public opinion when official mechanisms, such as the Letters and Visits System [*xinfang zhidu* 信访制度][3] and the mass media, are relatively ineffectual in monitoring and containing social injustice and official abuses of power. Online *weiguan* is affordable and practically achievable for most Chinese people equipped with digital devices and the Internet. It consists of various Internet-based micro actions, such as posting, reposting, commenting and sharing. These micro actions can rapidly generate large-scale communication networks in the online space around specific hot topics, putting pressure on the government to resolve controversial issues in the offline world.

Online *weiguan* differs in several respects from its offline predecessors. First, online *weiguan* is a transformative crowd activity. In the West, crowd activity has usually been taken as a synonym for social protest, and has been regarded as having the potential to create social and political change (Brighenti 2010; McClelland 1989; Rudé 1964). As Rudé (1964) succinctly put it, crowd activities should be viewed as an integral part of the social process and as such lend themselves to the exploration of popular politics. He suggests that studies of crowds should focus on activities, such as strikes, riots, rebellions and revolutions, which involve "aggressive mob", "hostile outburst" and "political demonstration" (Rudé 1964: 4). Accordingly, most offline *weiguan* phenomena in China should not be considered to be transformative crowd activity, because they seldom involve popular politics or advocate social and political change. They are either motivated by curiosity and a crowd mentality, or manipulated by political power to realize certain political ends. Only in cyberspace has *weiguan* become a form of popular politics and achieved political change on behalf of the people.

Like its offline antecedents, online *weiguan* also makes use of the "show-and-shame" ritual to mobilize public emotion against selected targets. However, the targets of contemporary *weiguan* are no longer selected by the CCP; instead, they are chosen by ordinary people. Moreover, online *weiguan* targets are usually CCP government officials, who are suspected of being involved in immoral and/or corrupt scandalous affairs. The ritual wielded by the CCP to punish its political opponents and manifest its disciplinary power has been reconfigured by disgruntled citizens to fight official corruption and human rights violations under the governance of the CCP. The collective power of crowds has become a double-edged sword, which can not only assist the CCP to fulfill its political purposes offline, but can also equip citizens to exercise supervisory power over the CCP's governance.

Second, like offline *weiguan*, online *weiguan* needs a gathering crowd. However, the connectivity, interactivity and networking of the Internet have made virtual crowd gathering possible. Therefore, online *weiguan* theoretically enables more heterogeneous participants to take part. Anyone who has a digital device and Internet access can participate without being restricted by temporal and spatial factors. As Naughton (2001) writes, by making use of digital networking devices, the "transaction costs" to organize, mobilize and participate in collection actions have been significantly reduced. In online *weiguan*, participants are not organized and manipulated by the CCP as in previous "shame-and-show" ceremonies.[4] Instead, they get together through Internet-enabled networking, mobilization and collaboration, based on their own interests and voluntary participation. This virtual, leaderless and networked *weiguan* action has become one of the most important collective actions in present-day China. It enables citizens to express disagreement, grievances and dissent and form public opinion in an environment in which open critique of the government and public gatherings are still tightly controlled and highly risky.

Third, online *weiguan* is mostly based on discursive communication with few corresponding offline actions. As a popular mode of online activism in China, it is also a "radical communicative action conducted in words and images" (Yang 2008). Symbolic and textual representations, such as verbal or visual storytelling, commentaries, satire and discussion, are the major discursive practices of online *weiguan*. A complete online *weiguan* process is normally divided into three communicative phases. The first phase is the storytelling of controversial issues, depicting wrongdoers and victims. The stories are told by victims, eyewitnesses or journalists, and posted online in the forms of words, images or videos. These stories usually have eye-catching titles, which underline the conflicts between the rich and poor, powerful and powerless. The second phase is emotional mobilization and the concentration of public opinion. The attention-grabbing posts are widely circulated and elicit strong emotional responses. These highly-mobilized public emotions, featuring sympathy toward victims and indignation at evils, can generate strong online public sentiment that advocates justice for victims and punishment for wrongdoers. The third phase is the "online-offline" and "bottom-up" flow of *weiguan* agendas. The issues publicized through *weiguan* can attract the attention of the mainstream media and government for further investigation. The online crowd usually disperses after government authorities respond to the online outcries and promise further investigations.

Online weiguan platforms: From BBS to micro-blog

Online *weiguan* is conducted chiefly on three platforms in China's cyberspace, namely BBS, HFS engines and micro-blogs. BBS was invented in the United States in the early 1970s (Jones 2003). It was first introduced to Chinese universities in the mid-1990s and soon enjoyed huge success among China's young people (Pan, Ling and Yu 2007). One of the earliest and most famous BBS sites was "Shuimu qinghua" [水木清华] established in 1995 at Tsinghua University. After 1998, BBS developed rapidly in China, and came to include such popular BBS sites as "Xici Hutong" [西祠胡同], "Strong Nation Forum"[*Qiangguo luntan* 强国论坛], "Tianya Community"[*Tianya shequ* 天涯社区], "Kdnet" [*Kaidi shequ* 凯迪社区], and "Baidu Tieba" [百度贴吧]. These BBS forums, consisting of hundreds of discussion boards and millions of threads and individual posts, continue to attract a large number of registered users; in June 2014, the number of China's BBS users reached 124 million (CNNIC 2014). BBS sites open up virtual communities for Internet users to share information and opinion on a broad spectrum of topics, particularly on hot-button issues. They have created an alternative space for the political engagement and social bonding of ordinary people in China's tightly-controlled and heavily-censored media environment (Yang 2003).

Many influential Internet incidents in the pre-blogging era were caused by *weiguan* on BBS sites. The cultural metaphor "construct a tall building" [*gai dalou* 盖大楼], which is widely used on BBS to refer to the prolonging of a conversation thread, is the predominant form of online *weiguan* on BBS sites. Though the population of BBS users has declined in recent years due to the popularity of blogs and micro-blogs, BBS is still playing an important role in online entertainment, discussion and participation (Hu 2010b).

HFS engines are another important online *weiguan* platform. Human Flesh Search [*Renrou sousuo* 人肉搜索] is a cyber-term invented by Chinese netizens. It refers to strategic problem-solving through online crowd sourcing, and aims to track down offline individuals by making an open appeal to the online community (Pan 2010; Herold 2011). HFS is not uniquely Chinese; similar phenomena exist in other social and cultural contexts. In *Here Comes Everybody: the Power of Organizing without Organization* (Shirky 2008), Shirky begins by telling an HFS-like story that occurred in New York. A lady lost her phone in a taxi. After a few days, the lady was able to find an e-mail address of the person who had picked up her phone, because this person had used her phone to post pictures she had taken a few days earlier. When the owner asked this person to return the phone, the current holder responded with "no". This set in motion a series of events in which netizens were able to use social networks to track down the culprit and bring her to the attention of the New York Police Department. Though similar examples of netizen vigilante justice have taken place in many countries, only Chinese netizens have embraced this cyber vigilantism as a regular practice to punish a wide range of people labeled "evil".

In China, mop.com [*maopu wang* 猫扑网] hosts the largest and most popular HFS engine. This was also China's first HFS engine, established in 2001 for users to submit and respond to requests for information. The mop site describes its HFS engine as "using collective knowledge instead of technology to solve problems" (Pan 2010). In the ensuing years, several major portal sites, such as Baidu, Sina and Yahoo, set up similar "question-and-answer" (Q&A) search engines. In contrast to major web search engines such as google.com, which scan their databases to retrieve sites related to the inputted keywords, HFS engines rely upon the collective labor of Internet users. They operate by enlisting as many people as possible to collaborate and contribute, in order to track down the individuals whom the original posting concerns. This collaborative and networked method is likely to attract high online traffic and give rise to online *weiguan*, and thus generates many Internet incidents. "Human flesh" [*renrou* 人肉] has become another popular cultural metaphor that vividly illustrates the practice of online *weiguan*.

BBS and HFS engines both lie somewhere between Web 1.0 and Web 2.0. They work on the same "forum and editor" model (Hu 2010b). With the development of Web 2.0 technologies, social networking sites (SNS) have become popular globally. They allow registers to "construct a public or semipublic profile within a bounded system", "articulate a list of other users with whom they share a connection" and "view and traverse their list of connections and those made by others within the system" (Boyd and Ellison 2007). Since their introduction in 1997, SNS such as MySpace, Facebook and Twitter have attracted millions of users. However, SNS that are popular in the West have a limited market in mainland China, due to Chinese government blocking (Flumenbaum 2009; Kerr 2009). The restriction of Western SNS has resulted in the rise of home-grown and state-approved SNS, such as Facebook equivalent Renren Network [*Renren wang* 人人网], and Twitter equivalent Weibo [微博]. The popularity of Weibo in particular has elevated online *weiguan* to new heights (Teng 2012).

Weibo is the Chinese translation of "micro-blogging". By June 2014, China's total Weibo population was 275 million (CNNIC 2014). Both Twitter and Weibo allow their users to share messages up to 140 characters in length, but 140 Chinese characters can convey much more meaning than the same number of Roman letters. In addition, Weibo provides users with a wider range of services than Twitter, including message threading and the ability to comment directly on other users' posts (Sullivan 2013). Technologically, Weibo integrates the elements of BBS, blogs, instant messaging systems and audio-visual sites, making it a truly multifunctional media platform. As the Online Public Opinion Monitoring Centre based at the *People's Daily* reported in March 2011, "micro-blogs have shown most vividly the speed and breadth of information transmission on the Internet, and they rapidly transmit information on the Internet with a means of high efficiency" (as cited in Bandurski 2011). The technological advantages of Weibo and its rapid popularization have helped netizens to "publicize and express their discontent" with greater frequency, and have posed "a new challenge to the state's regime of information control" (Sullivan 2013: 24).

The total number of Weibo users fell from 330 million at its peak in June 2013 to 275 million in June 2014 (CNNIC 2013; CNNIC 2014), due largely to the surging popularity of the WeChat [*Weixin* 微信] service.[5] This coincided with the Chinese government tightening control on Weibo through anti-Internet rumor campaigns and crackdowns on "Big-V"[6] (Yang 2014). Weibo users have therefore become more cautious about their expressions and many have migrated to Weixin, making *weiguan* activities on Weibo less active and frequent than before. However, Weibo's status as the most popular *weiguan* platform remains unchallenged because: (1). Weixin allows users to add a maximum of five thousand people to their personal networks, but Weibo users can follow or be followed by an unlimited number of people. (2). Weixin users' posts are only visible to a defined circle of accepted friends, whereas Weibo users' posts can be viewed by all followers and can be reposted and commented on with few limitations. Therefore, the semiprivate Weixin has much less potential to become a public forum than Weibo, and is not conducive to online *weiguan*, which require large-scale networking. Particularly in online events, Weibo remains strong and is irreplaceable (Wang and Fan 2014).

Platforms for online *weiguan* have expanded from BBS, to HFS engines, to the latest Weibo with the development of Web 2.0 technologies. However, the emergence and popularity of a new *weiguan* platform does not mean that the previous platforms have perished. A HFS posting may be reposted on BBS and attract a crowd on both platforms. A scoop concerning a scandal originating from Weibo may soon become a buzz topic on BBS and prompt HFS on the labeled "evil". Therefore, the three online *weiguan* platforms are not isolated, but interconnect in holding online *weiguan*. They work together with other forms of networked communication such as blogs, e-mail groups and instant chatting systems, increasing both the visibility of controversial issues, and engagement with *weiguan* activities.

Weiguan on BBS, HFS engines and Weibo: Three case studies

Hide-and-seek incident

On February 8, 2009, Li Qiaoming, a 24-year-old prisoner, was admitted to hospital from Jinning Detention Centre in Yunnan Province, and died four days later. On February 13, local newspaper *Yunnan Information Daily* first reported the incident and quoted the explanation of local police that Li died from a brain injury incurred when he ran into a wall while playing "duomaomao" [躲猫猫; a game similar to "hide-and-seek"] with his cellmates. The news was immediately reported by major portal sites in China, such as *Sina.com*, *Tencent.com* and *163.com*. Major BBS sites, personal weblogs and e-mail groups instantly jumped on the official explanation and questioned its veracity, and thus the incident quickly became a *weiguan* issue in cyberspace. Netizens suspected the local police had covered up the real cause of Li's death and urged the mainstream media and government to investigate.

Without convincing evidence to prove the local police's deceit, most netizens posted their online expressions in *e'gao* style. They created parodic comments with words and images, using mockery, sarcasm and irony to challenge the official explanation of the death of Li Qiaoming. On the Tianya Community BBS, netizens created posts under new ID names that included the term "duomaomao", such as "duomaomao youxi" [躲猫猫游戏; hide-and-seek game], and "dajia doulai duomaomao" [大家都来躲猫猫; let's play hide-

Illustration 1 The caricature shows three prisoners taking part in a hide-and-seek competition held in a cell; all hit their heads. (*News.qq.com* 2009a)

and-seek] (*Southern Metropolitan Daily* 2009). Comments included: "If you have children at home, don't allow them to play hide-and-seek", "Cherish your life and stay away from hide-and-seek", "How can the prisoners play hide-and-seek in their cells? If they can, where should the police go and find them?" A user with the ID name "manca1222" drafted rules for hide-and-seek: "1. Do not play with the wall; otherwise, the wall will kill you when you are caught. 2. If you do want to play, please make sure you have enough safety equipment, such as helmet, bullet-proof jacket, and so on" (*Southern Metropolitan Daily* 2009). In addition to sarcastic comments, some caricatures produced by anonymous netizens were also widely circulated on BBS sites and personal weblogs (see *News.qq.com* 2009a; *People.com.cn* 2009).

In addition to online critiques, the incident also evoked offline action, albeit on a smaller scale. A blogger named "shisan hu" put out a call for music videos. He wrote two songs about the hide-and-seek incident and posted them on his blog. These songs borrowed the tunes of a children's song and a popular song which most Chinese people are quite familiar with. However, he substituted the original lyrics with sarcastic lyrics. He called on netizens to sing his songs, record their singing with a camera and upload these music videos online. His appeal circulated on major BBS sites and drew many responses from netizens (Shisan hu's blog, February 24, 2009).

The production, circulation and consumption of these playful but critical discourses constructed a "counter-hegemony" (Gramsci 1971), challenging the local police's unbelievable explanation. These online expressions resonated with the public, formed

Illustration 2 The caricature shows a man, who represents "the public" [*gongzhong* 公众], looking for the truth with his eyes covered. Many hands are guiding him. The caricature suggests that seeking truth in China is like playing hide-and-seek. (*People.com.cn* 2009)

strong online public opinion, and further had offline influences. In contrast with the Sun Zhigang case, online public opinion was formed prior to the traditional media's investigation in the hide-and-seek incident. In 2003, online public opinion played a supplementary role in enhancing the communicative effects of investigative journalism by the traditional media, whereas by 2009, it had become the leading force in setting the agendas of both traditional media and government.

As a response to the public's online chatter, on February 20, a week after the first media report of Li's death, the local Publicity Department started to follow up the incident. It recruited ordinary netizens to set up a committee, and allowed them to enter the prison, check police documents and investigate the incident. A report written by the committee was later released to the public (*Rednet.cn* 2009; *News.qq.com* 2009b). Though the committee was unable to get to the bottom of the incident, and the investigation was questioned by netizens as a "government show" (*News.souhu.com* 2009), it had at least had proved one thing: online *weiguan* and the strong public sentiment it generated could force the government to take response.

Investigative reporting by mainstream media did not play a leading role in these developments, but was still important in expanding the social influence of this incident from online to offline, and making it a national news event. After the netizens' cell investigation, *SMD* and CCTV's investigative news program *Law Online* [*Fazhi zaixian* 法制在线] commissioned in-depth reports. Their reporting emphasized both online public opinion concerning the incident and the local government's response, with a particular focus on the unprecedented netizen investigation. The mainstream media thus changed their identity from investigator to commentator by adjusting the time and perspective of their reporting. They did not start to examine the incident until online public opinion had reached its peak. In this way, their reporting was guaranteed to be of interest to the public. Moreover, their reporting focused on the tension between netizens and police rather than challenging the police directly, and therefore was less risky.

The mainstream media's follow-up reporting consolidated online public opinion, expanded the influence of the incident to a national scale, and drew it to the attention of China's highest level of government. Zhou Yongkang, former Secretary of the Central Political and Law Commission, issued a demand that the truth must be found by a certain deadline. On February 27, the Yunnan Provincial Government Information Office told a news conference that the investigation had determined that the victim of the incident, Li Qiaoming, had died from assault by bullies in the detention centre. The previous explanation given by local police had proved to be a lie. Related authorities in the Jinning County Public Security Bureau and the detention centre, as well as the police on duty, were either removed or punished.

The scandal concerning a human rights violation ended with a victory for the netizens. Upon reflection, however, their actions were limited. In the incident, netizens failed to provide powerful evidence to the judicial organs; it was the official investigation that finally uncovered the truth. The main function of online *weiguan*, in this case, was to draw attention to a potential scandal by pressuring the government. In other cases, however, netizens have collaborated and contributed to the collection of evidence with more dramatic results.

HFS *weiguan*: Luxury cigarette case

On December 10, 2008, Zhou Jiugeng, director of Jiangning

District Property Bureau in Nanjing, accepted media interviews and said he would punish real estate developers who sold houses below cost price. His remark immediately evoked online indignation from urban residents, who could not afford the skyrocketing housing prices and expected them to decrease. Zhou Jiugeng's speech was completely at odds with public opinion, and thus instantly made him a target of online critique. On December 11, the day after Zhou's media interview, an Internet user wrote a post on KDnet.com and called on netizens to track Zhou down. This post was soon reposted on HFS engines and other BBS sites, starting a collaborative HFS on Zhou Jiugeng.

On December 14, a netizen named "huage" wrote a post entitled "Look at Zhou's cigarette". He said: "I happened to find a photo of Zhou at a conference. What a good civil servant he is! One carton of his cigarettes is worth three months of a laid-off worker's income". He posted the photo below his comment as evidence (*News.163.com* 2008).

On December 15, a netizen with the ID name "cheyou007" uncovered more evidence of Zhou's luxury lifestyle from another

Illustration 3 The photo shows Zhou Jiugeng delivering a speech at a conference. His cigarette, on the table, is circled. A note appears above, saying "What brand is this cigarette? It is 9-5 *zhizun* (九五至尊), Nanjing Cigarette Company's finest. One carton costs 1500 yuan."

Illustration 4 The photo shows Zhou Jiugeng sitting in a conference. He is wearing a Vacheron Constantin watch on his left wrist and holding a lit "9-5 *zhizun*" cigarette in his right hand.

photo. He wrote a post entitled "Zhou Jiugeng smokes brand cigarette and wears brand watch". He pointed out that Zhou's Vacheron Constantin watch was worth over 100,000 yuan. He posted this photo as evidence below his comment (*News.163.com* 2008).

These two photos of Zhou were quickly circulated online and attracted lengthy criticism of his luxury lifestyle. Netizens questioned how Zhou could afford such expensive items as an ordinary CCP official who earned a moderate wage, and wondered if Zhou misused public funds for personal consumption. In the next few days, more evidence of Zhou's luxury lifestyle and corruption was exposed online. Somebody witnessed Zhou driving a Cadillac to work. Someone said that Zhou's younger brother was a real estate developer, and so Zhou had cracked down on other real estate developers in order to protect his brother's business. Though this gossip about Zhou's private life was not verified, it contributed to the collaborative HFS and its results. The crowd-sourced evidence forced the government to hold an official investigation into this allegedly-corrupt official. On December 29, Zhou was dismissed from his position; he was later sentenced to eleven years in prison for bribery (Moore 2009).

It took just eighteen days to topple the high-ranking government official. This would have been a mission impossible without the participation and collaboration of netizens. In the Internet era, any information people give away about themselves in photos, surveillance cameras and online communities dramatically increases their social visibility, particularly if they are public figures such as government officials and celebrities. Once recorded, any improper words or behaviors can be used as evidence to punish their wrongdoings. In the case of Zhou Jiugeng, the process of crowd sourcing evidence of corruption formed online *weiguan*, generating strong public opinion to impose pressure on the government to punish alleged crimes.

Some people believe that HFS provides an effective means to fight official corruption in China. However, others argue that they offer only "illusory victories" (Elegant 2008), because it is not possible for corruption to be effectively resolved without genuine improvement in China's official anti-corruption mechanisms. Some also worry that HFS might violate personal privacy and call for legislation to regulate or ban this cyber activism (Bu 2013). However, it cannot be denied that HFS is an alternative means to exercise popular surveillance and public opinion supervision when official anti-corruption mechanisms are relatively ineffective.

Since 2009, Weibo has quickly overtaken BBS sites and HFS engines in organizing virtual crowd gathering. Its technological advantages and large number of subscribers have made Weibo a more networked, interactive and collaborative platform for online *weiguan*.

Weibo weiguan: Wenzhou high-speed train crash

On July 23, 2011, two high-speed trains collided on a viaduct at Shuangyu near Wenzhou in Zhejiang Province. Six carriages were derailed, four of which fell off the viaduct. The crash killed 40 people and injured 192. After the accident, the government hastily oversaw rescue operations, ordered the burial of the derailed carriages, and issued directives to limit media coverage. This intervention immediately evoked public outrage online. On China's most popular micro-blogging site, Sina Weibo, millions of users reacted furiously to the government's irresponsible and non-transparent way of dealing with the accident. They posted, commented and reposted on aspects of the accident, such as the death toll and the real cause, forming a huge online *weiguan* community in Weibo sphere.

Passengers on the train first released information about the accident through their Weibo accounts. Their eyewitness tweets attracted millions of Weibo users' attention prior to reports from the mainstream media. At 20: 38 p.m., four minutes after the accident, Yuan Xiaoyuan, a passenger on the train, updated her Sina Weibo via mobile phone, saying: "Something has happened to D301 in Wenzhou. There was an emergency shutdown and a strong collision. Twice! The power is out. I am in the last carriage. I'm praying I'll be okay. It is too terrifying now". Another D301 passenger, named "yangjuan quanyang", called out for help thirteen minutes after the accident, writing on her Weibo account: "Help! The high-speed train D301 has derailed somewhere not far from Wenzhou South Station. Children are crying! There is no staff coming. Come and help us!" This SOS tweet was reposted more than 100,000 times in the space of minutes (A. Li 2011). From 20:38 p.m. to 23:57 p.m., Yuan Xiaoyuan sent out a total of seven tweets (Yuan Xiaoyuan's Weibo, July 23, 2011).

Compared with previous channels for self-styled reporting, such as personal blogs and audio-visual portals, Weibo supports users to update their accounts via mobile phones in a faster and more convenient way. Sina Weibo even provides a voicemail service to its users—in some emergency situations, when keypad typing on the mobile phone is not available, users can simply dial a service number on their mobile phones and record a message. The message is then automatically uploaded and shared (Ye 2011). The technological advantages of Weibo have made it the most important real-time media for crisis and emergency reporting since 2009. Moreover, Weibo has become an open platform for critiquing official information. In the Weibo sphere, professional journalists, public intellectuals, celebrities and ordinary citizens work together to challenge the official information released by government or mainstream media. They either share alternative news to prove the bias of official information, or produce critical comments in a rumoresque or parodic way.

In the aftermath of the train tragedy, a spokesperson for the

Ministry of Railways, Wang Yongping, said in an interview with CCTV that the crash was caused by bad weather. The train D3115 was hit by thunder and subsequently lost power; it then hit the moving D301 from the back. This explanation attracted immediate criticism. The public suspected that the Ministry of Railways had covered up the real cause of the crash and urged for the truth to be released. Another bone of contention came after the government ordered the burial of the train wreckage. On July 24, a citizen journalist uploaded a three-minute-long video clip to Youku.com that recorded bulldozers burying parts of the wreckage near the accident scene. At least two bodies appeared to fall out of the carriages during the burial. The link and snapshots of the video clip spread around Weibo. Netizens questioned why the search-and-rescue mission had been called off so quickly and why important evidence was hastily buried before careful examination. In the first press conference since the collision held by the Ministry of Railways on the evening of July 24, Wang Yongping responded to public questions. He said the burial was purely for the convenience of the emergency rescue operation, rather than to bury evidence. He added: "whether you believe it or not, I certainly do". When asked why rescuers had pulled a toddler out of the wreckage alive hours after the rescue effort had been officially called off, Wang said: "it was a miracle". Wang's arrogant tone and unconvincing answers soon became more fodder for criticism.

In the absence of credible information from the government and mainstream media, people tend to look for alternative news to fill the information gap. Weibo has provided such a platform. In the wake of the train crash, users who could access news beyond the Party-state-controlled Chinese media posted alternative news items on Weibo. These alternative news posts were quickly shared and commented on by a large number of users, forming sites for alternative news *weiguan*. Qian Gang, former editor of *Southern Weekend* and currently a research scholar based at the University of Hong Kong, updated his Weibo with snapshots of headlines about the rail tragedy from newspapers based in Hong Kong. On July 25 and 26, Qian wrote eight tweets to share the gist of critical reporting from Hong Kong newspapers, including *Apple Daily, Oriental Daily, Ming Pao, Wenhui Daily, AM 730* and *Headline Daily*. Each of his tweets was reposted an average of roughly five thousand times and received about one thousand comments (Qian Gang's Weibo, July 25–26, 2011). Similarly, Li Miao, a correspondent from Hong

Kong's Phoenix Satellite TV based in Tokyo, posted a video clip from a Phoenix Satellite news program at 19:31 on July 25. In this video, Li Miao interviewed a Japanese bullet train expert. Li succinctly summarized:

> Japanese bullet trains have met with thunder before, but no one died. The emergency response technology on China's high-speed trains must be defective . . . Rescue work and preserving the scene are highly important in railway accidents in Japan; the burial of the train wreckage is just ridiculous.

Li's word-video combined tweet had been posted 26,920 times and received 5,890 comments by the evening of July 26 (Li Miao's Weibo, July 26, 2011). Similar alternative news tweets attracted huge attention from Weibo users who doubted the official information but had no means to access alternative media outlets. In addition to alternative news tweets posted by information "have-more" groups—chiefly, professional journalists and public intellectuals—tweets generated by the information "have-less" grassroots also triggered online crowd gathering. These tweets were either written like "inside information", exposing secrets behind the official information, or written in a parodic way to mock the Ministry of Railways's cover-up. Both forms mobilized public sentiment to pressure the government to release the truth.

On July 24, the Ministry of Railways reported that thirty-five people had died. This Ministry of Railways toll was immediately questioned by speculation on Weibo:

> 35 is a miraculous number. The death toll of the Wenzhou high-speed train crash is 35. The death toll of the mining disaster in Pingdingshan, Henan Province, is 35. The death toll of the Chongqing rainstorm is 35. The death toll of Yunnan rainstorm is 35 . . . Do you know why the death toll is always less than 36? Because when the death toll passes 36, the Secretary of the Municipal Committee of the CCP will be removed from their post. Therefore, no matter what accident happens, the death toll is bound to be less than 36. (As cited in Mai Xiaomai's Weibo, July 25, 2011)

Another rumor tweet that was widely reposted concerned the disappearance of Chai Jing. It has been rumoured that Chai Jing, one of China's most famous investigative journalists, from CCTV's top investigative show, *News Probe*, was threatened and detained after she spoke out on her Weibo about her plan to investigate the train crash independently (*Maydaily.com* 2011). Millions of Weibo users reposted this tweet and left comments to praise Chai Jing's courage, support her investigation, and condemn the government's strict control of independent journalistic practice.

According to Shibutani, rumors emerge when people attempt to "construct a meaningful interpretation" of an ambiguous situation by "pooling their intellectual resources". Rumors should not be interpreted as reflecting the inaccuracy of reports transmitted by word-of-mouth, but rather as "a form of collective problem-solving" (1966: 17). Though the rumor tweets about the death toll and Chai Jing's disappearance were verified as faked news later, they "challenge[d] official reality by proposing other realities" and "[constrained] authorities to talk while contesting their status as the sole source authorized to speak" (Kapferer 1990: 215, 14). As Hu Yong (2009) argues, in China, where free speech is restricted and news production is censored, rumor regularly functions as social protest in emergency and crisis events. In the highly-networked and interactive new media environment, the rapid circulation and wide consumption of rumors can produce what Harsin (2006) calls a "rumor bomb", and develop sufficient force to check the powers that dominate the production and release of information.

Like rumors and citizen reporting, tweets written in *e'gao* style also attracted high online traffic. As previously mentioned, Wang Yongping, spokesman for the Ministry of Railways, became a target for criticism due to his arrogant way of talking to the public. The perfunctory answers he gave at the first press conference after the train crash ("Whether you believe it or not, I certainly do", "It's a miracle" and "I can only tell you it did happen") immediately provoked sarcastic critique from Weibo users. Netizens transplanted these three sentences into new contexts and produced new commentaries. This parodic writing style, dubbed "high-speed train style" [*gaotie ti* 高铁体], soon gained popularity online and inspired a "sentence-making" competition. Below is one of the most reposted "high-speed train style" tweets on Sina Weibo:

> A woman reported that a government official raped her. Government: "He wore a condom, so it wasn't rape". Woman: "He didn't wear a condom; he lied". Government: "Whether you believe it or not, I certainly do". Woman: "But I am pregnant now".

> Government: "It is a miracle". Woman: "How can you . . . ?" Government: "I can only tell you it did happen". (As cited in Ju Detao's Weibo, July 25, 2011)

Creative *e'gao* works produced by anonymous netizens were also widely reposted. In a blueprint for a "never colliding high-speed train", the designer placed several government officials in the first and last carriages, where they would be spared from the full impact of any train crash (see Figure 4.3). This satirized a common situation—that government officials enjoy higher-quality products and much better service than other people. If government officials regularly took the high-speed train, its designer reasoned, the quality and safety of the train would be greatly improved. In the movie poster, the designer Photoshopped pictures of Sheng Guangzu (the Party Chief and Minister of Railways), Wang Yongping, and Long Jing (the head of the Shanghai Railway Bureau) into the background of train wreckage. The three government officials were named as leading actors in a disaster film called *Fatal Bullet Train*. The top of

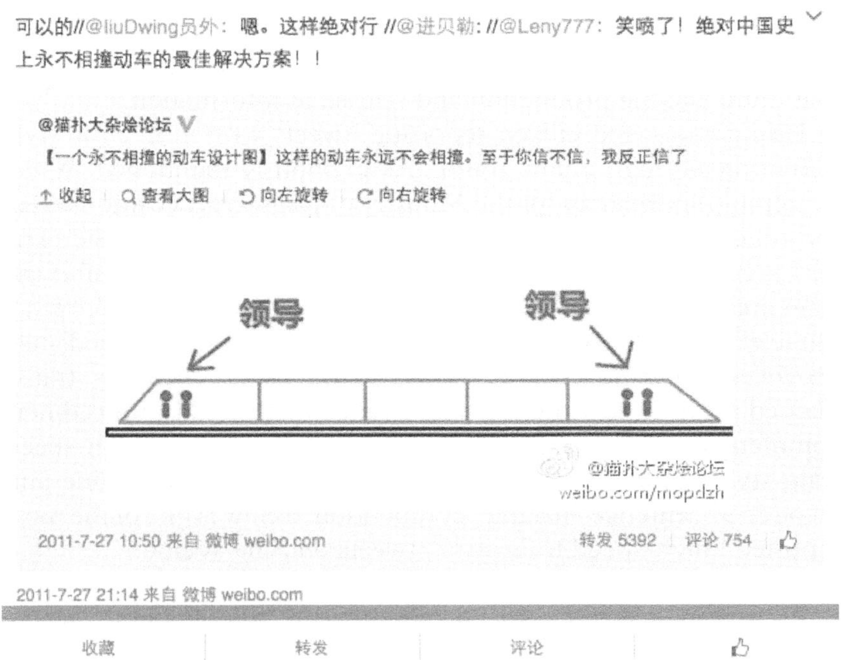

Figure 4.3 "Never colliding high-speed train" blueprint (Shanghai jimo xiaosheng's Weibo, July 27, 2011).

Figure 4.4 Poster of the film *Fatal Bullet Train* (Xiaodan tongxue qiu shunli's Weibo, July 28, 2011).

the poster indicated that the movie would be a blockbuster to celebrate the ninetieth anniversary of the founding of the CCP (see Figure 4.4).

As Regan (2003) put it, "creativity almost always happens at the edges of societies, not in the centre". The people deprived of "the right to know" after the accident used parody and social media platforms to create strategic expressions. In this process, they were empowered to express their outrage against the Ministry of Railways' cover-up in a playful way.

The eyewitness reporting tweets, alternative news tweets, rumor tweets and parody tweets constituted four major sites for *weiguan*, and attracted huge attention from Weibo users. They read, commented, posted and reposted, forming a networked *weiguan* community to exert pressure on the government to take action. As a result, this active spectatorship effectively checked and shaped official ways of dealing with the accident. The buried train wreckage was excavated and moved to Wenzhou Western Railway Station for further investigation on July 26. The government increased the compensation for victims and their families to a total of 91.5 million yuan—a new high for railway accidents (*Chinahourly.com* 2011). The Ministry of Railways finally admitted that faulty signal technology had led to the collision (*Reuters.com* 2011). On July 28, Chinese Premier Wen Jiabao headed to Wenzhou, where he visited the site of the train crash to pay his respects to the victims, comfort the injured, and promised a top-level investigation into the accident (Ramzy 2011b).

Summary

Online *weiguan* is an Internet-mobilized collective action with a long cultural and political history in Chinese society. It has provided unprecedented publicity for political scandals and has transformed the conventional management and reporting of controversial social issues. Through online *weiguan*, ordinary Chinese people have gained opportunities to mobilize public emotion, form public opinion and generate group pressure for the government to respond to scandals. At a time when both citizens' direct participation in public affairs and the supervisory function of the mass media are difficult to achieve, the failures of official democracy have led to a burgeoning unofficial democracy in the "third realm" (Huang

1993) between China's state and society. Online *weiguan* is one avenue for realizing this unofficial democracy. Its purpose, practice and ethos have all had political implications for Chinese society.

In its supervisory capacity, online *weiguan* has greatly challenged Foucault's Panopticon model of a disciplinary society (1977) which allows the powerful few to monitor the powerless many through the normalizing power of the gaze. In the online society, people are transformed from objects under scrutiny into proactive spectators who gaze upon "the few", forming what Mathiesen (1997) terms "a synoptic system of power". Cascio (2005) colorfully calls this power structure a "Participatory Panopticon", in which citizens equipped with digital devices carry out constant surveillance through networked communication.

Online *weiguan* allows ordinary Chinese people to exercise the critical practices of "active spectatorship" (Kreiss, Meadows and Remensperger 2015). This active spectatorship is usually practiced in the form of mediated scrutiny through social media representations rather than physical spectatorship offline, and has become synonymous with "collective participation" in China's heavily-controlled speech environment. It has shifted previous intensive and high-risk social movements into more decentralized, extensive and flexible online protests, based on netizens' "opt-in/opt-out" (Bennett and Segerberg 2012) coalitions and the "cultural logic of networking" (Juris 2005). Online *weiguan* has generated a new form of political participation and has constituted an important part of what Guobin Yang (2009a) calls "China's new citizen activism". This new citizen activism, embedded in the everyday use of social media, cultivates Chinese people's sense of citizenship and reinforces the notion that making a better China is an ongoing commitment that requires every individual's participation, contribution and collaboration. As popular sayings in China's cyberspace go, "attention is power" [*guanzhu jiushi liliang* 关注就是力量] and "spectatorship changes China" [*weiguan gaibian zhongguo* 围观改变中国].

5

Internet Interventionism and Deliberative Politics in China's Web 2.0 Era

Alternative and activist uses of the Internet have provided opportunities and possibilities for challenging and transforming different types of media events via agents, agendas, performances and political impacts. As a result, media events are no longer exclusively represented by mainstream media and controlled by the Party and the market. Instead, they have been widely represented and discussed by ordinary people through a variety of new media platforms. The decentralization of media events has formed a dynamic "discursive event 'sphere'" (Volkmer 2008: 90), in which state and non-state players compete, negotiate and interact to direct the narrative structures of events, articulate their own agendas and realize different political purposes.

The case studies in this book have demonstrated that online activism has created alternative channels for ordinary people to participate in politics informally and voice their grievances, demands and priorities to the government. In the meantime, it has opened up unofficial channels for the state to listen to public needs, detect popular undercurrents and co-opt social resistance. Accordingly, the state has adjusted its practices surrounding both media events specifically and everyday governance more generally, so as to maintain its own legitimacy. Online activism has thus shaped the interactions and interrelations between the public and the state in China, and is forming a new political trend of "Internet interventionism".

Internet Interventionism and the Mediation of Politics

Internet interventionism refers to Internet-enabled political engagement, through which multiple social actors communicate, interact, and adapt to each other and effect social and political change through strategic use of the Internet. I propose this concept based on the broader concept of "media interventionism", which Jesper Strömbäck and Frank Esser argue is "a media-centered political reporting style in which, increasingly, journalists and media actors become the stories' main newsmakers rather than politicians or other social actors" (2009: 217). Media interventionism is a kind of journalistic intervention, which emphasizes the crucial role of journalistic practices in defining and constructing political realities. As Strömbäck and Esser succinctly explain the interventional mechanism, "the media logic trumps political logic with respect to how the media cover politics" (2009: 216). This journalistic interventionism can force political actors and institutions to accommodate and adapt to media logic (Cook 2005; Strömbäck 2008).

This theory was developed to explain the interventional role of media in the West. It is difficult to conceive of applying it in China, where media and journalistic practices are tightly controlled by the Party-state, and so media logic must frequently give in to political logic or commercial logic. Such was the case in the Chinese investigative journalism discussed in Chapter 4. However, the popularization of the Internet since the mid-1990s has decentralized the media power controlled by the Party-market, and has created participatory channels and platforms for multiple social actors to claim media power. In the Internet era, media power is unlikely to be monopolized by mainstream media and their "power-money". Instead, it has become a contested terrain in which multiple social actors compete to intervene in social, cultural and political issues. The rise of alternative and activist uses of the Internet has facilitated the formation of "Internet interventionism", which goes beyond conventional journalistic intervention in several important respects.

First, Internet interventionism has greatly extended social agency in media intervention. Media intervention is no longer the exclusive domain of political elites and professional journalists; everyone who has access to the Internet can participate in such interventional practice and exercise media power. They can be grassroots entrepreneurs, citizen journalists, public intellectuals, social activists, or other types of ordinary citizens. These non-state social actors take the Internet as a representational tool which affords them the space to express and organize themselves in ways that parody, modify, complement or challenge the dominant and institutionalized ways of "doing" culture and politics in Chinese society.

Second, Internet interventionism has diversified the practice of media interventionism such that it now goes beyond mere "political reporting". It has generated innovative practices to intervene in politics in direct and indirect ways. Thus, *shanzhai* media culture challenges the dual monopolies of the Chinese media in ideology and market by spoofing and copycatting media spectacles; citizen journalism complements or subverts the mainstream media's crisis reporting; online *weiguan* exercises popular surveillance and brings the pressure of public opinion to bear on the government by organizing, mobilizing and networking millions of netizens to post, repost and comment on scandalous events. These practices, embedded in the everyday Internet communications and consumption of Chinese people, seek social and political changes in non-official and sometimes non-serious ways. They come from multiple social, cultural and political perspectives, making media intervention more affordable, flexible, frequent and extensive.

Consequently, Internet interventionism has greatly increased the mediation of politics in scope and degree. Mediation, as Silverstone argues, is "a fundamentally dialectal notion which requires us to address the processes of communication as both institutionally and technologically driven and embedded" (2005: 189). The concept provides an insightful perspective through which to examine the "interdependence and mutual shaping of communication action and communication technology" (Lievrouw 2011: 234). Following this approach, the mediation of politics refers to political communicative actions which utilize media technologies. On the one hand, the media have become a relatively independent institution with a logic of their own that political actors in other institutions must adapt to. On the other, the media have been incorporated by different political actors into the working mechanisms of their own institutions and have functioned to serve various political purposes.

The mediation of politics is not a new phenomenon in the Internet era. As Finnemann argues, "[n]ew media—whenever they arrive in history—imply extended mediatisation" (2008: 15). Not only the Internet but also other "old" media, such as telegraph, telephone and television, have all extended existing communicative forms and have transformed patterns of social, cultural and political interaction. However, for the first time in history, the Internet allows information (texts, graphics, sound and audio) to be rapidly and effectively transmitted across multiple media platforms. This has caused not only the digitization of media and communication technologies (Fidler 1997) but also the convergence of new and old media (Jenkins 2006). In the Internet-centric, digitized and converged media environment, the patterns of political communication have been innovated, reformed and transformed. The mediation of politics has accordingly increased.

Non-state players increasingly depend on the Internet to obtain alternative information, articulate social critiques and organize political activities, taking the Internet as a "microphone" and "loudspeaker". However, state powers increasingly rely on the Internet to listen to people's voices, respond to public sentiment and resistance, and enhance governing capability, seeing the Internet as a "safety valve". The Internet has thus become an interventional tool that multiple social actors use to pursue different political agendas. Internet interventionism should not be viewed as a one-sided process in which non-state players mediate politics from the bottom up; the reverse process should also be acknowledged, as state powers mediate politics from the top down.

The bi-directional nature of Internet-enabled political communication has created new forms of interaction between the state and non-state in China. The preceding case studies have shown that the intervention of online activism has forced the state to reform its traditional ways of handling media events and politics. CCTV has incorporated audiences' online and offline participation into its annual Spring Festival Gala, and transformed the gala into a more open and interactive media spectacle. The government has increasingly deployed new media in managing disasters and political scandal events in order to calm public sentiment and guide online public opinion. For example, when Weibo became a popular online *weiguan* platform to mobilize for collective action and advocacy, it was swiftly incorporated by the government into everyday publicity work and governance to neutralize public opinion and promote interaction with citizens between crises. By the end of 2014, the number of government-affiliated Weibo accounts had reached 130,103. Of these, 94,164 were run by government institutions and 35,939 were run by government officers (*News.xinhuanet.com* 2015). By the end of June 2014, government-held Sina Weibo accounts had generated 24.35 million posts, which in turn had been reposted and commented on 317 million times (*News.sina.com.cn* 2014). Integrating Weibo into everyday governance has reflected the CCP's pursuit of a more responsive style of governance and has demonstrated the ability of China's one-Party state to adapt in the digital era (Noesselt 2014).

This digitally-mediated interaction between the state and the non-state has facilitated a deliberative turn in China's political development. For non-state players, this deliberation has democratic potential. As exemplified in the preceding chapters, they promote discursive participation and citizen engagement through online activism practices, creating popular and unofficial forms of debate even as authentic "deliberative democracy" (Cohen 1997; Dahlgren 2009) through rational debate and consensual decision-making cannot be realized in China. The deliberation practiced by the Party-state is more authoritarian. Central and local government has proactively and strategically used the Internet and other new media to negotiate with the public and enhance their public management capabilities. Grassroots voices and sentiments have unprecedented opportunities to be heard by the government and influence the policy-making process. However, whether the expansion of "listen-to" channels can ensure effective "listen-in" results remains questionable. The government's adaptive use of the Internet aims primarily to establish a more transparent, responsible and quasi-democratic image of the Chinese government, rather than actually democratizing China's authoritarian political system. This phenomenon constitutes what He and Warren (2011) call "authoritarian deliberation", through which the authoritarian regime incorporates deliberative practices into its governance to stabilize and strengthen its rule. Although these deliberative acts seemingly transform authoritarian governance and sometimes empower society, they exclusively promote what Warren (2009) calls "governance-driven democratization".

According to Warren (2009), the trend of governance-driven democratization exists mostly in developed democracies, but can also be seen in authoritarian societies such as China, because of similar market-driven neoliberal development and the increasing strength and pluralism of Chinese society. The striking feature of this innovative practice is that it has created deliberative opportunities to facilitate interactions between state power and civil society (Warren 2009). Hence, it has enabled the government to capture the potential within civil society for organization, information, energy and creativity, and has thus made administrative powers more responsive and efficient (Hajer and Wagenaar 2003; Warren 2009). This trend is a kind of political experiment, which may have democratic potential but will not lead to "regime-level democratization" (Warren 2009: 4).

Digitally-mediated political communication has gradually become an effective political strategy for both the state and non-state players, and has formed a deliberative politics with Chinese characteristics. This politics features "talk-centric" public deliberation (Chambers 2003) in the virtual space with few offline corresponding actions, and takes Internet interventionism as a fundamental mechanism to further social and political change. It promotes progressive and small-scale social adaptations and arrangements, rather than adopting violent revolution and pushing drastic political reform. In the paradoxical process of deliberation, the state and society, public and private, structure and action, material and symbolic, online and offline dynamically interplay. The case studies in the book have demonstrated the variety of digitally-mediated political communications as well as the deliberative politics in China's Web 2.0 era.

New Horizons for the study of China's online activism

Of course, much remains to be explored in the field of Chinese online activism beyond the book. Consideration of practices such as hacking, for example, may illuminate more radical aspects of online activist culture as well as the official responses to it. Similarly, while the present book has investigated the interventions of online activism within China's borders, transnational online activism may prove a fruitful line of inquiry to examine the overseas Chinese democracy movements and China's transnational civil society in the new media era. In addition to considering hitherto unexplored online activism practices, future research might also address non-democratic aspects of online activism. For example, HFS, a popular way of exercising online *weiguan*, has facilitated Chinese people's supervision of official power abuses in various political scandals. However, if deployed in cyber-violence, HFS could raise alarming legal issues, such as slander, defamation and violations of privacy law. Also, online activism usually articulates popular sentiment that promotes the nationalistic ideology of the CCP, enhancing the legitimacy of the state rather than contributing to democratic debate.

The governance of China's online activism is similarly rich in opportunities for discussion. With the development and evolution of Internet technologies, the state is continually updating its censorship and regulations on the Internet to cope with various online activism practices. Beyond such technological and legal control, market forces have also been used to curb public sentiments fuelled by online activism. For example, different levels of government recruit propagandist Internet commentators, who are popularly dubbed the "50 Cent Party" by Chinese netizens. These government agents post comments online to promote pro-government opinion, particularly when government institutions and officials are forced to defend their credibility. The "Internet Water Army" [*wangluo shuijun* 网络水军], a group of Internet ghost-writers who are hired by private public relations companies and paid to post comments with particular content, plays a similar role in balancing online public opinion. The participation of these pseudo-grassroots players has made it difficult to verify the authenticity and credibility of voices and opinions from the real "masses" [*remmin qunzhong* 人民群众] in cyberspace.

The rise of online activism has forced the state to reform and innovate its governance of the Internet. Accordingly, practitioners of online activism are constantly developing new strategies to elude, fight against or adapt to state regulations. The mutual constitution and co-evolution of online activism and Internet governance has made the study of China's Internet more complex and challenging. More in-depth empirical research is needed to investigate the transformations, practices and functions of online activism as well as the social, cultural and political factors that influence it. It will also be important to compare and contrast China's online activism with that of other (semi-) authoritarian states and democratic nations to understand the commonality and uniqueness of Chinese online activism. By examining China's evolving online activism and the mediated social and political change it evokes, the dynamics of the interplay between China's state and society will continue to unfold.

Glossary

Baidu tieba	百度贴吧	http://tieba.baidu.com/index.html
baijia jiangtan	百家讲坛	Lecture Room
Beijing zhichun	北京之春	Beijing Spring
chaoji nüsheng	超级女声	Super Girls
chunjie lianhuan wanhui	春节联欢晚会	Spring Festival Gala
daju yishi	大局意识	sense of the big picture
dangxing yuanze	党性原则	Party principle
dazibao	大字报	big-character poster
Duli jilupian yundong	独立纪录片运动	Independent Documentary Movement
duomaomao	躲猫猫	hide-and-seek
e'gao zhifu	恶搞之父	the father of *e'gao*
Fazhi zaixian	法制在线	Law Online
gai dalou	盖大楼	"construct a tall building"
gaotie ti	高铁体	high-speed train style (a satirical writing style)
gei zuguo muqin baidanian	给祖国母亲拜大年	wish the Motherland a Happy Spring Festival
Gongmin diaocha	公民调查	Citizen Investigation
gongzhong	公众	the public
guanzhu jiushi liliang	关注就是力量	attention is power
Guanzhu wenchuan dizhen	关注汶川地震	Focusing on the Wenchuan Earthquake
hexie shehui	和谐社会	harmonious society

Huaer weishenme zheyang hong	花儿为什么这样红	Why Are the Flowers So Red?
hulianwang disu zhifeng	互联网低俗之风	wave of online smut
Huwen xinzheng	胡温新政	Hu-Wen New Deal
Jiaoban yangsh, gei quanguo renmin bainian	叫板央视，给全国人民拜年	Challenge CCTV and wish all Chinese a happy Spring Festival
Jiaodian fangtan	焦点访谈	Focus Interview
Jintian	今天	Today
Kaidi shequ	凯迪社区	http://www.kdnet.net
kanke wenhua	看客文化	culture of gaze
kongsu	控诉	denunciation
Lao ma ti hua	老妈蹄花	Disturbing the Peace (Ai Weiwei's documentary on Tan Zuoren case)
Liang Shan Bo	梁山泊	[a name of a place in Outlaws of Marshes]
maopu wang	猫扑网	http://www.mop.com
Minjian chunwan	民间春晚	Folk Spring Festival Gala
minjian kanwu	民间刊物	non-governmental periodicals
Muzimei xianxiang	木子美现象	Muzimei phenomenon
Nanfang dushibao	南方都市报	Southern Metropolitan Daily
Nanfang zhoumo	南方周末	Southern Weekend
Nian	念	Remembrance
Niubo wan	牛博网	www.bullogger.com
piping yu ziwo piping	批评与自我批评	criticism-self criticism
Qiangguo luntan	强国论坛	http://bbs1.people.com.cn
Qiu Jin	秋瑾	(a female pioneer of the 1911 Revolution)
Renmin chunwan renmin ban, banhao chunwan wei renmin	人民春晚人民办，办好春晚为人民	A People's Gala Held by the People, for the People
renmin qunzhong	人民群众	the masses
Renren wang	人人网	http://www.renren.com

renrou sousuo	人肉搜索	human flesh search engine
Shanzhai chunwan	山寨春晚	*Shanzhai* Spring Festival Gala
Shanzhai nian	山寨年	the Year of *Shanzhai*
Shimin yuban shanzhaiban chunwan jiaoban yangshi	市民欲办山寨版春晚叫板央视	Ordinary citizen intends to run *shanzhai* version of Spring Festival Gala to challenge CCTV
Shuihu zhuan	水浒传	Outlaws of the Marsh
Shuimu qinghua	水木清华	http://www.newsmth.net
sixiang gongzuo	思想工作	thought work
suku	诉苦	speaking bitterness
Tianya shequ	天涯社区	http://www.tianya.cn
tonggao	通稿	template
Wangchuan	忘川	Forgetting Sichuan
wangluo shijian	网络事件	Internet incident
wangluo shuijun	网络水军	Internet Water Army
wangluo weiguan	网络围观	online surrounding gaze
wangluo yulunnian	网络舆论年	the year of online public opinion
Weibo	微博	micro-blog
weiguan	围观	surrounding gaze
weiguan gaibian zhongguo	围观改变中国	spectatorship changes China
wei renmin fuwu	为人民服务	serve the people
Weixin	微信	WeChat
Women de wawa	我们的娃娃	Our Children
Woyao shang chunwan	我要上春晚	I Want to Perform on the Spring Festival Gala
Wuji	无极	The Promise
wumao dang	五毛党	50 Cent Party
Wusi luntan	五四论坛	May Fourth Forum
Wuyier xuesheng dang'an	5.12 学生档案	5.12 Student Archive
Xici Hutong	西祠胡同	http://www.xici.net
Xidan nühai	西单女孩	(a subway female singer)
Xin chunjie wenhua xuanyan	新春节文化宣言	New Spring Festival Culture Manifesto
xinfang zhidu	信访制度	Letters and Visits System
Xin jilupian yundong	新纪录片运动	New Documentary movement

xin meiti shijian	新媒体事件	new media events
xin minsu	新民俗	new folk custom
Xinwen lianbo	新闻联播	Network News
Xuri yanggang	旭日阳刚	(a migrant worker duo)
yanda	严打	campaigns against criminal activities
Yao	药	Medicine (novella by Lu Xun)
yidi jiandu	异地监督	extra-regional media supervision
Yige mantou yinfa de xue'an	一个馒头引发的血案	The Bloody Case of a Steamed Bun
yindao yulun	引导舆论	guide public opinion
yulun jiandu	舆论监督	supervision by public opinion
Zhongguo fazhi baodao	中国法制报道	The Chinese Law Report
Zhongguo renquan	中国人权	China Human Rights
Zhongguo shanzhai dianshitai	中国山寨电视台	China Countryside Television
zuiniu dingzihu	最牛钉子户	the coolest nail-house
zuoda zuoqiang	做大做强	make [China's media industry] bigger and stronger

Notes

Introduction

1 The term "non-state" refers to individuals, groups or organizations that are not strictly controlled by the state. It does not belong to or exist as a state-structure or established institution of a state, but has sufficient power to influence and cause changes in politics, such as the social activists, non-governmental organizations (NGOs) and alternative media outlets. The relations between the state and the non-state actors could be confrontational or collaborative (see Qian and Xu 1993; Yu 2009).

2 Admittedly, the media reform era and the new media era do not have clear-cut boundaries and partially overlap. China officially came online in 1994, when the commercialization and marketization of the nation's media was well underway. Similarly, China's media reforms began in 1978, but continue to develop with new technologies. Nevertheless, these categories are analytically useful in understanding the broad trends within Chinese media over recent decades.

1 Alternative Media and Online Activism

1 The case of Sun Zhigang is one of the earliest and most significant Internet incidents in China. Sun was a 27-year-old university graduate and fashion designer from Wuhan who worked in Guangzhou. He was detained because he could not produce his temporary living permit and ID card during a routine inspection by Guangzhou police. Three days later, on March 20, 2003, Sun died in the detention centre from police violence. His death was soon reported by the *Southern Metropolitan Daily* [*Nanfang dushibao* 南方都市报] (*SMD*) and provoked heated online discussion. Due to the strength of public pressure shaped by online activism, the government investigated Sun's death and two people subsequently received the death penalty. More importantly, this case led to the abolishment of China's detention and repatriation system (see Zhao 2008a; Yu 2009).

2 Media Celebration: *Shanzhai* Media Culture as Media Intervention

1 "Thought work" refers to the CCP's ideological propaganda and

political persuasion (Lynch 1999). It is "the main means for guaranteeing the party's ongoing legitimacy and hold on power" (Brady 2006: 59), and has been "the very life blood of the Party-state" (Brady 2008: 1) in the post-1989 era. According to Brady (2008), the CCP's thought work has evolved with China's changing social, cultural, political and technological environment in order to maintain the Party's ideological control and governing legitimacy.

3 Media Disaster: Citizen Journalism as Alternative Crisis Communication

1 In the Chinese context, "liberal media" refer to media outlets that are greatly influenced by the style of professional journalism and liberal and democratic orientation of the West. The liberal media have a tradition of investigative journalism and underscore their role in exercising public opinion supervision. To call these media outlets "liberal" does not mean that they are free from the CCP's regulation and censorship. Rather, the term emphasizes their difference in content and function to other traditional mainstream media that work as apparatus for propaganda. The most famous liberal media in China are the *Southern Weekend* [*Nanfang zhoumo* 南方周末] and *SMD*, which are both owned by the Southern Daily Group based in Guangdong Province.

2 Details of Ai Weiwei's citizen investigation project in this and subsequent paragraphs were obtained from http://aiweiwei.com/wp-content/uploads/2013/12/公民调查事件回顾（2008年05月12日－2010年05月12日）.pdf(accessed May 20, 2014).

3 Bullogger.com ceased to operate on July 3, 2013, but. Ai Weiwei's blog entries concerning his citizen investigation can be found on his personal homepage http: //aiweiwei.com (accessed April 20, 2014).

4 Media Scandal: Online *Weiguan* as Networked Collective Action

1 In popular Chinese parlance, "flies" refer to low-ranking officials and "tigers" refer to high-level corrupt officials.

2 "Extra-regional media supervision" is also translated as "cross-region media supervision". In China, investigative reporting by the media in any given region is likely to be censored by the local publicity department. By investigating politically unfavorable news in other regions, local media can avoid regional restrictions and establish a national reputation. "Extra-regional supervision" is thus an effective tactic for investigative reporting in China's hierarchical and authoritarian press system (see Cho 2010).

3 The "Letters and Visits System" was established in the PRC in the 1950s as an official channel for Chinese citizens to seek assistance from government to resolve their grievances. In the absence of just and

transparent legal and judiciary systems in China, the Letters and Visits System is one of the main methods for ordinary citizens to seek institutional recourse to redress violations of legal rights, property rights, and other human rights (see Minzner 2006).

4 The participants of online *weiguan* are mainly voluntary and unorganized ordinary citizens. However, members of "50 Cent Party" (*Wu mao dang* 五毛党) who are organized and manipulated by the CCP often participate in the process as well. "50 Cent Party" refers to state-sponsored online commentators. They widely post comments favourable towards the CCP's policies on BBS, chatrooms, blogs and Weibo to shape and sway public opinion, particularly when government organs and officials are forced to defend their credibility in controversial social issues. It is said the commentators are paid 50 cents *yuan* (1/2 yuan) per posting.

5 WeChat, also known as "Weixin", literally translates as "micro message" in English. It is a mobile messaging app (similar to WhatsApp) released by Tencent, the company behind the "QQ" instant messaging service, in January 2011. It provides hold-to-talk voice messaging, text messaging, broadcast (one-to-many) messaging and photo and video sharing. By the first quarter of 2014, WeChat had 549 million monthly active users in China and overseas (Statista 2015).

6 "Big-V" is the moniker for verified accounts on Weibo, who are opinion leaders with millions of followers in the Weibo sphere. The government's "Big-V" crackdown aims to discipline them and minimize their online authority and influence. Xue Manzi, a "Big-V" with more than 12 million followers who is known for his critical postings on controversial social issues in China, was arrested in Beijing on August 23, 2013, accused of having sex with a prostitute (Buckley 2013).

Bibliography

Ai, X. 2008: On film, not as art but as propaganda and as agent for change. *ASIANetwork Exchange* XV (3), 8 & 21.

Ai, X. 2010, May 12: Women de wawa 我们的娃娃 (Our Children). Retrieved from http://www.youtube.com/watch?v=YKzAdoqsz3U

Ai, X. 2010, July 13 and 14: Wan shexiangji de nürenbang 玩摄像机的女人帮 (The women who use cameras). Retrieved from http://antichinagfw.blogspot.com.au/2010/10/gfw-blog_2238.html

Alexander, D. 2005: An interpretation of disasters in terms of changes in culture, society and international relations. In R.W. Perry and E.L. Quarantelli (eds.), *What is a Disaster? New Answers to Old Questions*, United States:Xlibris Corporation, 25–38.

Allan, S. 2002: Reweaving the internet: Online news of September 11. In B. Zelizer and S. Allan (eds.), *Journalism After September 11*, London and New York: Routledge, 119–40.

Allan, S. 2007: Citizen journalism and the rise of "mass self-communication": Reporting the London bombings. *Global Media Journal* (Australian Edition) 1 (1), 1–20.

Ang, A. 2009, May 7: China—5,335 students dead, missing in 2008 quake. Retrieved from http://www.postbulletin.com/china-students-dead-missing-in-quake/article_128d67c3-9b06-5b7f-a7f5-8ecd83dd42fc.html

Askci.com 2009: 2003–2008 nian zhongguo shouji yonghu shuliang ji zengzhang 2003–2008 年中国手机用户数量及增长 (Number and growth of mobile phone users in China, 2003–2008). Retrieved from http://www.askci.com/data/viewdata54096.html

Atton, C. 1999: A re-assessment of the alternative press. *Media, Culture and Society* 21 (1), 51–76.

Atton, C. 2003: Reshaping social movement media for a new millennium. *Social Movement Studies* 2(1), 3–15.

Atton, C. 2004: *An Alternative Internet: Radical Media, Politics and Creativity*. Edinburgh: Edinburgh University Press.

Bakhtin, M. 1984: *Rabelais and His World* (H. Iswolsky, Trans). Bloomington: Indiana University Press.

Bandurski, D. 2011, October 3: Can social media push change in China? Retrieved from http://cmp.hku.hk/2011/10/03/15870/

Baoye.net 2009, January 13: Guizhou dianshitai: Yinshi erdong yinshi erbian 贵州电视台：因势而动因时而变 (Guizhou TV Station develops with changing media environment). Retrieved from http://www.baoye.net/News.aspx?ID=285865

Beijingreview.com.cn 2009, April 3: Shanzhai wenhua shi chuangxin haishi qinquan? 山寨文化是创新还是侵权 (Is Shanzhai culture an innovation or infringement?). Retrieved from http://www.beijingreview. com.cn/jd/txt/2009-04/03/content_189550.htm

Bennett, W. L. and Segerberg, A. 2012: The logic of connective action: Digital media and the personalization of contentious politics. *Information, Communication & Society* 15 (5), 739–68.

Berry, C. 2003: The documentary production process as a counter-public: Notes on an inter-Asian mode and the example of Kim Dong-Won. *Inter-Asia Cultural Studies* 4 (1), 139–44.

Berry, C. 2007: Getting real: Chinese documentaries, Chinese postsocialism. In Z. Zhang (ed.), *The Urban Generation: Chinese Cinema and Society at the Turn of the Twenty-first Century,* Durham and London: Duke University Press, 115–34.

Bivens, R. and Li, C. 2010: Web-oriented public participation in contemporary China. In S. Tunney and G. Monaghan (eds.), *Web Journalism: A New Form of Citizenship*? Brighton: Sussex Academic Press, 275–88.

Boin, A. 2005: From crisis to disaster: Towards an integrative perspective. In R.W. Perry and E.L. Quarantelli (eds.), *What is a Disaster? New Answers to Old Questions*, United States: Xlibris Corporation, 153–72.

Boyd, D.M. and Ellison, N.B. 2007: Social network sites: Definition, history, and scholarship. *Journal of Computer–Mediated Communication* 13: 210–30.

Brady, A.M. 2006: Guiding hand: The role of the CCP Central Propaganda Department in the current era. *Westminster Papers in Communication and Culture* 3 (1), 58–77.

Brady, A.M. 2008: *Marketing Dictatorship: Propaganda and Thought Work in Contemporary China.* Lanham, MD: Rowman & Littlefield.

Branigan, T. 2011, June 23: Artist Ai Weiwei released on bail by Chinese police. Retrieved from http://www.theguardian.com/artanddesign/2011/jun/22/ai-weiwei-freed-by-chinese-police

Brighenti, A.M. 2010: Tarde, Canetti, and Deleuze on crowds and packs. *Journal of Classical Sociology* 10 (4), 291–314.

Bu, Q. 2013: "Human flesh search" in China: The double-edged sword. *International Data Privacy Law* 3 (3), 181–96.

Buckley, C. 2013, September 10: Crackdown on bloggers is mounted by China. Retrieved from http://www.nytimes.com/2013/09/11/world/asia/china-cracks-down-on-online-opinion-makers.html

Cammaerts, B. 2007: Jamming the political: Beyond counter-hegemonic practices. *Continuum: Journal of Media & Cultural Studies* 21 (1), 71–90.

Cao, P. 2010: *Media Incidents: Power Negotiation on Mass Media in Time of China's Social Transition*. Konstanz: Uvk.

Carducci, V. 2006: Culture jamming: A sociological perspective. *Journal of Consumer Culture* 6 (1), 116–38.

Cascio, J. 2005, May 4: The rise of the participatory Panopticon. Retrieved from http://www.worldchanging.com/archives/002651.html

Castells, M. 2000: *The Rise of the Network Society* (2nd ed.). Oxford; Malden, MA: Blackwell.

Castells, M. 2007: Communication, power and counter-power in the networked society. *International Journal of Communication* 1 (1), 238–66.

Chadwick, A. 2006: *Internet Politics: State, Citizens, and New Communication Technologies*. New York and Oxford: Oxford University Press.

Chambers, S. 2003: Deliberative democracy theory. *Annual Review of Political Science* 6, 307–26.

Chan, Y. 2010: The journalism tradition. In D. Bandurski and M. Hala (eds.), *Investigative Journalism in China: Eight Cases in Chinese Watchdog Journalism*, Hong Kong: Hong Kong University Press, 1–17.

Chase, M. S. and Mulvenon, J. C. 2002: *You've Got Dissent! Chinese Dissident Use of the Internet and Beijing's Counter-Strategies*. Santa Monica, CA: Rand Corporation.

Cheek, T. 2006: Xu Jilin and the thought work of China's public intellectuals. *The China Quarterly* 186, 401–20.

Chen, C. 1996, September 17: University students transmit messages on defending the Diaoyu islands through the Internet, and the authorities are shocked at this and order the strengthening of control. *Sing Tao Jih Pao*. In FBIS, September 18, 1996.

Chen, L. 2008: Open information system and crisis communication in China. *Chinese Journal of Communication* 1 (1), 38–45.

Cheung, A.S.Y. 2007: Public opinion supervision: A case study of media freedom in China. *Columbia Journal of Asian Law* 20 (2), 357–84.

Chinahourly.com 2011, November 19: Wenzhou motor car accident: 19 families of the victims have signed compensation agreements. Retrieved from http://www.chinahourly.com/bizchina/201111/30691.html

China Media Project 2011: The surrounding gaze 围观. Retrieved from http://cmp.hku.hk/2011/01/04/9399/

Cho, L. F. 2010: The origins of investigative journalism: The emergence of China's watchdog reporting. In D. Bandurski and M. Hala (eds.), *Investigative Journalism in China: Eight Cases in Chinese Watchdog Journalism*, Hong Kong: Hong Kong University Press, 165–76.

CNN.com 2004, January 6: Country breakdown: Probable cases of SARS. Retrieved from http://edition.cnn.com/2003/HEALTH/05/28/sars.breakdown/

CNNIC 2003: Di 12 ci zhongguo hulian wangluo fazhan zhuangkuang tongji baogao 第12次中国互联网络发展状况统计报告 (The 12th statistical report on Internet development in China). Retrieved from http://news.xinhuanet.com/ziliao/2003-07/22/content_986928.htm

CNNIC 2008: Di 22 ci zhongguo hulian wangluo fazhan zhuangkuang tongji baogao 第22次中国互联网络发展状况统计报告 (The 22nd statistical report on Internet development in China). Retrieved from http://download.xinhuanet.com/it/document/cnnic22.pdf

CNNIC 2013: Di 32 ci zhongguo hulian wangluo fazhan zhuangkuang tongji baogao 第32次中国互联网络发展状况统计报告 (The 32nd statistical report on Internet development in China). Retrieved from http://www.cnnic.cn/hlwfzyj/hlwxzbg/hlwtjbg/201307/P02013071750 5343100851.pdf

CNNIC 2014: Di 34 ci zhongguo hulian wangluo fazhan zhuangkuang tongji baogao第 34 次中国互联网络发展状况统计报告 The 34th statistical report on Internet development in China). http://www.cnnic.cn/hlwfzyj/hlwxzbg/hlwtjbg/201407/P02014072150 7223212132.pdf

CNNIC 2015: Di 36 ci zhongguo hulian wangluo fazhan zhuangkuang tongji baogao第 36 次中国互联网络发展状况统计报告 The 36th statistical report on Internet development in China). Retrieved from: http://www.cac.gov.cn/files/pdf/hlwtjbg/hlwlfzzktjbg036.pdf

Cohen, J. 1989: Deliberation and democratic legitimacy. In A. Hamlin and P. Pettit (eds.), *The Good Polity: Normative Analysis of the* State, Oxford: Basil Blackwell, 17–34.

Cohen, J. 1997: Deliberation and democratic legitimacy. In J. Bohman & W. Rehg (eds.), *Deliberative Democracy: Essays on Reason and Politics*, Cambridge: The MIT Press, 67–92.

Cohiba.blogcn.com 2005, August 28: Yangshi youbian lao'er, pinpin chuzhao daya chaonü 央视忧变老二，频频出招打压超女 (CCTV worries about losing dominant status, cracks down *Super Girl*). Retrieved from http://cohiba.blogcn.com/articles/央视忧变老二-频频出招打压超女-转-2.html

Cook, T. E. 2005: *Governing with the News: The News Media as a Political Institution* (2nd ed.). Chicago: University of Chicago Press.

Coombs, W. T. 1998: The Internet as potential equalizer: New leverage for confronting social irresponsibility. *Public Relations Review* 24 (3), 289–303.

Coombs, W. T. 2007: *Ongoing Crisis Communication: Planning, Managing, and Responding* (2nd ed.). Thousand Oaks, CA: Sage.

Coppola, D. P. 2011: *Introduction to International Disaster Management* (2nd ed.). Boston: Butterworth-Heinemann Ltd.

Cottle, S. 2006: Mediatized rituals: Beyond manufacturing consent. *Media, Culture and Society* 30 (1), 411–32.

Cottle, S. 2011: Media and the Arab uprisings of 2011. In S. Cottle and L. Lester (eds.), *Transnational Protests and the Media*, New York: Peter Lang, 293–304.

Couldry, N. 2003: *Media Rituals: A Critical Approach*. London: Routledge.

Couldry, N. 2004: Theorizing media as practice. *Social Semiotics* 14 (2), 115–32.

Couldry, N. and Curran, J. (eds.) 2003: *Contesting Media Power: Alternative Media in a Networked World*. Lanham, MD: Rowman and Littlefield.

Dahlgren, P. 2009: *Media and Political Engagement: Citizens, Communication, and Democracy*. New York: Cambridge University Press.

Dai, J. 2007: Incorporating the resistance? A case study on the appropriation of the *Promise* (*Wu Ji*). Paper presented at the annual conference of the International Communication Association (ICA), May 24–28, San Francisco.

Dai, X. 2000: Chinese politics of the Internet: control and anti-control. *Cambridge Review of International Affairs* 13 (2), 181–94.

Dartnell, M. 2006: *Insurgency Online: Web Activism and Global Conflict*. Toronto: University of Toronto Press.

Dayan, D. 2008: Beyond media events: Disenchantment, derailment, disruption. In M. E. Price and D. Dayan (eds.), *Owning the Olympics: Narratives of the New China*, Ann Arbor, MI: Michigan University Press, 391–401.

Dayan, D. and Katz, E. 1988: Articulating consensus: The ritual and rhetoric of media events. In J. C. Alexander (ed.), *Durkheimian Sociology: Cultural Studies*, Cambridge: Cambridge University Press 161–86.

Dayan, D. and Katz, E. 1992: *Media Events: The Live Broadcasting of History*. Cambridge, MA: Harvard University Press.

De Burgh, H. 2003: Kings without crowns? The re-emergence of investigative journalism in China. *Media, Culture & Society* 25 (6), 801–20.

Dery, M. 1993: *Culture Jamming: Hacking, Slashing and Sniping in the Empire of Signs*. New York: Open Magazine Pamphlet Series.

Deuze, M. 2003: The Web and its journalisms: Considering the consequences of different types of newsmedia online. *New Media & Society* 5(2), 203–30.

Deuze, M. 2007: *Media Work*. Cambridge: Polity Press.

DiMaggio, P., Hargittai, E., Neuman, W. R. and Robinson, J. P. 2001: Social implications of the Internet. *Annual Review of Sociology* 27 (1), 307–36.

Downing, J., Ford, T., Gil, G. and Stein, L. 2001: *Radical Media: Rebellious Communication and Social Movements*. London: Sage Publications.

Elegant, S. 2008, December 31: Exposing corruption on the Internet:

Illusory victories? Retrieved from http://china.blogs.time.com/2008/12/31/exposing-corruption-on-the-internet-illusory-victories/

Ent.qq.com 2011, November 11: Shanzhai chunwan laomeng chu xinzhao 山寨春晚老孟出新招 (Shanzhai Spring Festival Gala's Lao Meng has a new idea). Retrieved from http://ent.qq.com/a/20111111/000257.htm

Fan, M. 2008, May 18: Chinese media take firm stand on openness about earthquake. Retrieved from http://www.washingtonpost.com/wp-dyn/content/article/2008/05/17/AR2008051701790.html

Fearn-Banks, K. 1996: *Crisis Communications: A Casebook Approach.* Mahwah, NJ: Erlbaum Associates.

Fidler, R. 1997: *Mediamorphosis: Understanding New Media.* Thousand Oaks, CA: Pine Forge Press.

Finnemann, N. O. 2008: The Internet and the emergence of a new matrix of media: Mediatization and the coevolution of old and new media. Paper presented at the AoIR Internet 9.0, October 15–18, Copenhagen. Retrieved from http://zh.scribd.com/doc/38225571/The-Internet-and-the-Emergence-of-Aoir-Conf

Fisk, J. 1994: *Media Matters: Everyday Culture and Political Change.* Minneapolis: University of Minnesota Press.

Flew, T. 2005: *New Media: An Introduction.* Oxford: Oxford University Press.

Flumenbaum, D. 2009, July 3: China blocks twitter ahead of Tiananmen anniversary. Retrieved from http://www.huffingtonpost.com/2009/06/02/china-blocks-twitter-ahea_n_210177.html

Foucault, M. 1977: *Discipline and Punish: The Birth of the Prison* (A. Sheridan, Trans.). Harmondsworth: Penguin.

Fraser, N. 1990: Rethinking the public sphere: A contribution to the critique of actually existing democracy. *Social Text* 25/26, 56–80.

Fritz, C. 1961: Disaster. In R. Merton and R. Nisbet (eds.), *Contemporary Social Problems.* NY: Harcourt, 651–94.

Fuchs, C. 2010: Alternative media as critical media. *European Journal of Social Theory* 13 (2): 173–92.

Fuchs, C. 2011: *Foundations of Critical Media and Information Studies.* New York: Routledge.

Garcia, D. and Lovink, G. 1997, May 16: The ABC of tactical media. Retrieved from http://www.nettime.org/Lists-Archives/nettime-l-9705/msg00096.html

Giddens, A. 1991: *Modernity and Self-Identity: Self and Society in the Late-modern Age.* Stanford, CA: Stanford University Press.

Gillette, S., Taylor, J., Chavez, D., Hodgson, R., and Downing, J. 2007: Citizen journalism in a time of crisis: Lessons from California wildfires. *The Electronic Journal of Communication* 17 (3 & 4). Retrieved from http://www.cios.org/EJCPUBLIC/017/3/01731.HTML

Gillmor, D. 2006: *We the Media: Grassroots Journalism by the People, for the People*. Sebastopol, CA: O'Reilly Media.

Glasser, T. L. and Ettema, J. S. 1989: Investigative journalism and the moral ordee. *Critical Studies in Media Communication* 6 (1), 1–20.

Glazer, M. 2006, September 27: Your guide to citizen journalism. Retrieved from http://www.pbs.org/mediashift/2006/09/your-guide-to-citizen-journalism270.html

Gong, H. and Yang, X. 2010: Digitized parody: The politics of *e'gao* in contemporary China. *China Information* 24 (1), 3–26.

Gonzales-Herrero, A. and Pratt, C. B. 1996: An integrated symmetrical model for crisis communication management. *Journal of Public Relations Research* 8 (2), 79–105.

Good, K. 2006: The rise of the citizen journalist. *Feliciter* 52 (2), 69–71.

Gramsci, A. 1971: *Selections from the Prison Notebooks* (Q. Hoare and G. Nowell-Smith Trans.). London: Lawrence & Wishart.

Grossberg, L. 1992: *We Gotta get out of this Place: Popular Conversation and Postmodern Culture*. New York and London: Routledge.

Grossberg, L. 1993: Cultural studies and/in new worlds. *Critical Studies in Mass Communication* 10, 1–22.

Gu, C. 2008, November 20: Huangyan tizhixia de ziyou zhuyi meiti kuozhan 谎言体制下的自由主义媒体扩展 (The development of liberal media under China's deceptive system). Retrieved from http://www.chinesepen.org/Article/sxsy/200811/Article_20081120145524.shtml

Habermas, J. 1989: *The Structural Transformation of the Public Sphere: An Inquiry into a Category of Bourgeois Society* (T. Burger and F. Lawrence, Trans.). Cambridge: Polity Press.

Haddow, G. D., Bullock, J. A. and Coppola, D. P. 2007: *Introduction to Emergency Management*. Boston: Butterworth-Heinemann Ltd.

Hajer, M. A. and Wagenaar, H. (eds.) 2003: *Deliberative Policy Analysis: Understanding Governance in the Network Society*. Cambridge: Cambridge University Press.

Hall, S. 1980: Encoding/decoding. In S. Hall, D. Hobson, A. Lowe, and P. Willis (eds.), *Culture, Media, Language*. London: Hutchinson, 128–40.

Hall, S. 1986: On postmodernism and articulation: An interview with Stuart Hall (L. Grossberg, ed.). *Journal of Communication Inquiry* 10 (2), 45–60.

Harrison, M. 2002: Satellite and cable platforms: Development and content. In S. H. Donald, M. Keane and H, Yin (eds.), *Media in China: Consumption, Content and Crisis*. London: RoutledgeCurzon, 167–78.

Harsin, J. 2006: The rumor bomb: Theorizing the convergence of new and old trends in mediated US politics. *Southern Review* 39 (1), 85–110.

Harvey, D. 2005: *A Brief History of Neoliberalism*. Oxford: Oxford

University Press. He, B. and Warren, M. E. 2011: Authoritarian deliberation: The deliberative turn on Chinese political development. *Perspectives on Politics* 9 (2), 269–89.

He, Q., Li, J. and Wu, W. 2009, December 3: Wenchuan dizhen baodao, wangluo buru zhuliu meiti 汶川地震报道，网络步入主流媒体 (Internet has become mainstream media in reporting on Sichuan earthquake). Retrieved from http://www.hljnews.cn/by_xwcb/system/2009/12/03/010503078.shtml

He, Z. 1998: Cong houshe dao dangying yulun gongsi: Zhonggong dangbao de yanhua 从喉舌到党营舆论公司：中共党报的演化 (From mouthpiece to Party Publicity Inc: The evolution of the Chinese Communist Party's press). In Z. He & H. Chen (eds.), *Zhongguo Chuanmei Xinlun*中国传媒新论 (*New Perspectives on the Chinese Media*), Hong Kong: Taipingyang shiji chubanshe 太平洋世纪出版社, 66–107.

Hebdige, D. 1979: *Subculture: The Meaning of Style*. London and New York: Routledge.

Hepp, A. 2004: *Netzwerke der Medien: Medienkulturen und Globalisierung* (Networks of the media: Media cultures and globalization). Wiesbaden: VS.

Hepp, A. and Couldry, N. 2010: Introduction: Media events in globalized media cultures. In N. Couldry, A. Hepp, and F. Krotz (eds.), *Media Events in a Global Age*, London and New York: Routledge, 1–20.

Herold, D. K. 2011: Human flesh search engines: Carnivalesque riots as components of a "Chinese democracy". In D. K. Herold, and P. Marolt (eds.), *Online Society in China: Creating, Celebrating and Instrumentalising the Online Carnival*. London and New York: Routledge, 127–45.

Hewitt, K. 1983: The idea of calamity in a technocratic age. In K. Hewitt (ed.), *Interpretations of calamity: From viewpoint of human ecology*, Boston, MA, London and Sydney: Allen and Unwin, 3–32.

Ho, J. 2010: *Shanzhai*: Economic /cultural production through the cracks of globalization. Retrieved from http://sex.ncu.edu.tw/members/Ho/20100617%20Crossroads%20Plenary%20Speech.pdf

Hu, Y. (2009): Yaoyan zuowei yizhong shehui kangyi 谣言作为一种社会抗议(Rumour as a kind of social protest). *Chuanbo yu shehui jikan* 传播与社会季刊 (*Communication and Society Quarterly*), 9, 67–94.

Hu, Y. 2010a, November 25: Weibo, kanke ruhe shixian luodi? 微博，看客如何实现落地？ (How do microbloggers act?) Retrieved from http://21ccom.net/articles/sxpl/pl/article_2010112525254.html

Hu, Y. 2010b, June 1: BBS sites on China's changing web. Retrived from http://cmp.hku.hk/2010/06/01/6158/

Huang, C. P. 1993: "Public sphere"/ "civil society" in China? The third realm between state and society. *Modern China* 19 (2), 216–40.

Huang, Q. 2006, July 22: Parody can help people ease work pressure.

Retrieved from http://www.chinadaily.com.cn/cndy/2006-07/22/content_646887.htm

Hutzler, C. 2010, March 26: China bans poet from travelling to US conference. Retrieved from http://archives.truthaboutchina.com/2010/03/china-bans-poet-from-traveling.html

Ido.3mt.com.cn 2009, January 22: Shanzhai chunwan jiduochou: Yanchu changdi bei quxiao, wangluo zhibo cheng paoying 山寨春晚几多愁：演出场地被取消，网络直播成泡影 (Shanzhai Spring Festival Gala's troubles: Venue cancelled and no online broadcast). Retrieved from http://ido.3mt.com.cn/Article/200901/show1258008c26p1.html

Ifeng.com 2010, February 9: Yangshi chunwan guanggao shouru chao liudianwu yi 央视春晚广告收入超6.5亿 (Advertising income from CCTV Spring Festival Gala exceeds 650 million yuan). Retrieved from http://finance.ifeng.com/money/roll/20100209/1814263.shtml

Jacobsson, K. and Lofmarck, E. 2008: A sociology of scandal and moral transgression: The Swedish "Nannygate" scandal. *Acta Sociologica* 51 (3), 203–16.

Jenkins, H. 2006: *Convergence Culture: Where Old and New Media Collide.* New York: New York University Press.

Jiang, F. 2009: Game between "quan" and "shi": Communication strategy for Shanzhai subculture in China cyberspace. Retrieved from http://www.scribd.com/doc/15919031/Fei-Jiang-Chinese-Shanzhai-Culture-Studies

Jiang, M. 2012: Chinese internet events. In A. Esarey and R. Kluver (eds.), *The Internet in China: Online Business, Information Distribution and Social Connectivity*, New York: Berkshire Publishing.

Jiang, W. and Ma, J. 2009: Shanzhai chunwan de feizhengchang siwang 山寨春晚的非正常死亡 (The unnatural death of Shanzhai Spring Festival Gala). *Qingnian Zhoumo* 青年周末 (*Youth Weekend*) 3. Retrieved from http://www.dooland.com/magazine/article_1961.html

Jiang, Y. 2014: "Reversed agenda-setting effects" in China: Case studies of Weibo trending topics and the effects on state-owned media in China. *The Journal of International Communication* 20 (2): 168–83.

Jones, S. (ed.) 2003: *Encyclopaedia of New Media: An Essential Reference to Communication and Technology*. Thousand Oaks, CA: Sage.

Jordan, T. 2002: *Activism! Direct Action, Hacktivism and the Future of Society*. London: Reaktion Books.

Ju Detao's Weibo (2011, July 25). Retrieved from http://weibo.com/1951152870

Juris, J. S. 2005: The new digital media and activist networking within anti-corporate globalization movements. *Annals of the American Academy of Political and Social Science* 597 (1), 189–208.

Kao, H. and Lee, J. 2010: The application of shanzhai innovation model in China: The examples of mobile phone, notebook computer, and automobile. Paper presented at the Summer Conference on "Opening Up Innovation: Strategy, Organization and Technology", June 16–18, London. Retrieved from http://www2.druid.dk/conferences/viewpaper.php?id=500846&cf=43

Kapferer, J. 1990: *Rumors: Uses, Interpretations, and Images*. New Brunswick, NJ: Transaction.

Katz, E. and Liebes, T. 2007: 'No more peace'! How disaster, terror and war have upstaged media events. *International Journal of Communication* 1, 157–66.

Keane, M. and Zhao, J. 2012: Renegades on the frontier of innovation: The shanzhai grassroots communities of Shenzhen in China's creative economy. *Eurasian Geography and Economics* 53 (2), 216–30.

Kellner, D. 1997: Intellectuals, the public sphere, and new technologies. *Research in Philosophy and Technology* 16, 15–32.

Kellner, D. 2003: *Media spectacle*. London and New York: Routledge.

Kellner, D. and Share, J. 2007: Critical media literacy, democracy, and the reconstruction of education. In D. Macedo and S. R. Steinberg (eds.), *Media Literacy: A Reader*. New York: Peter Lang Publishing, 3–23.

Kerr, R. 2009, July 7: China blocks Twitter, Facebook, most Web 2.0: Google and YouTube also censored as a means of hiding bloody riots. Retrieved from http://vator.tv/news/2009-07-07-china-blocks-twitter-facebook-most-web-20

Kidd, D. 1999: The value of alternative media. *Peace Review* 11 (1), 113–19.

Kidd, D. 2003: Indymedia. Org: A new communications commons. In M. McCaughey and M. D. Ayers (eds.), *Cyberactivism: Online Activism in Theory and Practice*, New York and London: Routledge, 47–69.

Kreiss, D., Meadows, L., and Remensperger, J. 2015: Political performance, boundary spaces, and active spectatorship: Media production at the 2012 Democratic National Convention. *Journalism* 16 (5), 577–95.

Lam, O. 2009, February 5: Shanzhai Spring Festival Gala blocked in China. Retrieved from http://advocacy.globalvoicesonline.org/2009/02/05/shanzhai-spring-festival-gala-blocked-in-china/

Langlois, A. and Dubois, F. (eds.) 2005: *Autonomous Media: Activating Resistance and Dissent*. Montreal: Cumulus Press.

Lao Meng's blog (2008, November 23): Shanzhai chunwan zhengji jiemu 山寨春晚征集节目 (*Shanzhai* Spring Festival gala calls for programs). Retrieved from http://blog.sina.com.cn/s/blog_045571650100brs8.html

Lasn, K. 1999: *Culture Jam: The Uncooling of America* ™. New York: Eagle Brook.

Lee, C. C., Chan, J., Pan, Z. and So, C. 2002: *Global Media Spectacle: News War Over Hong Kong*. Albany: State University of New York Press.

Lee, C., He, Z. and Huang, Y. 2006: 'Chinese Party Publicity Inc.' conglomerated: The case of the Shenzhen Press Group. *Media, Culture & Society* 28 (4), 581–602.

Leng, X. and Zhang, M. 2011: Shanzhai as a weak brand in contemporary China marketing. *International Journal of China Marketing* 1 (2), 81–94.

Leonard, A. 1999, October 8: Open-source journalism. Retrieved from http://www.salon.com/1999/10/08/geek_journalism/

Lévi-Strauss, C. 1966: *The Savage Mind*. Chicago: University of Chicago Press.

Li, A. 2011, August 16: China's micro-blogs break the ban. Retrieved from http://www.dc4mf.org/en/node/412

Li, H. 2011: Parody and resistance on the Chinese Internet. In D. K. Herold and P. Marolt (eds.), *Online Society in China: Creating, Celebrating and Instrumentalising the Online Carnival*, London and New York: Routledge, 71–88.

Li, J. 2009: Qiantan gonggong weijizhong de meiti gongneng yu zeren 浅谈公共危机中的媒体功能与责任 (Media's function and responsibility in public crisis). Retrieved from http://www.studa.net/Movie/100723/1634102.html

Li Miao's Weibo (2011, July 26). Retrieved from http://weibo.com/limiao-tokyo

Lievrouw, L. A. 1994: Information resources and democracy: Understanding the paradox. *Journal of the American Society for Information Science* 45 (6), 350–57.

Lievrouw, L. A. 2011: *Alternative and Activist New Media*. Cambridge and Malden: Polity Press.

Lin, Y. J. 2011: *Fake Stuff: China and the Rise of Counterfeit Goods*. London and New York: Routledge.

Liu, A. P. 1971: *Communications and National Integration in Communist China*. Berkley, CA: University of California Press.

Lu, X. [1919] 1960: Medicine. In Lu X. (ed.), *Selected Stories of Lu Hsun* (H. Yang and G. Yang, Trans.). Beijing: Foreign Language Press.

Lü, X. 2003a: Jiedu 2002 nian chunjie lianhuan wanhui 解读2002年春节联欢晚会 (Reading CCTV's 2002 Spring Festival Gala). Retrieved from http://www.cssm.gov.cn/view.php?id=28038

Lü, X. 2003b: *Jilu Zhongguo: Dangdai Zhongguo Jilupian Yundong* 记录中国: 当代中国纪录片运动 (*Recording China: The Contemporary Chinese Documentary Movement*). Beijing: Sanlian shudian 三联书店.

Lü, X. 2009: Ritual, television, and state ideology: Rereading CCTV's 2006 Spring Festival Gala. In Y. Zhu and C. Berry (eds.), *TV China*, Bloomington: Indiana University Press, 111–25.

Lull, J. and Hinerman, S. 1997: The search for scandal. In J. Lull and S. Hinerman (eds.), *Media Scandals: Morality and Desire in the Popular*

Culture Marketplace. New York: Columbia University Press, 1–33.

Lynch, D. C. 1999: Dilemmas of "thought work" in *fin-de-siècle* China. *The China Quarterly* 157, 173–201.

Ma, J. and Zhou, P. 2009: Shijue sheyun: Ai Xiaoming Bu Wei duitan 视觉社运: 艾晓明、卜卫对谈 (Visual social movement: dialogue between Ai Xiaoming and Bu Wei). *Chuanbo yu shehui jikan* 传播与社会季刊 (*Communication and Society* Quarterly) 10, 197–212.

MacKinnon, R. 2010, July 16: China's Internet censorship and controls: The context of Google's approach in China. Retrieved from http://www.hrichina.org/content/3248

Mai Xiaomai's Weibo (2011, July 25). Retrieved from http://weibo.com/karolin

Markovits, A. S. and Silverstein, M. (eds.) 1988: *The Politics of Scandal: Power and Process in Liberal Democracies*. New York: Holmes & Meier.

Marra, F. J. 1998: The importance of communication in excellent crisis management. *Australian Journal of Emergency Management* 13 (3), 6–8.

Mathiesen, T. 1997: The viewer society: Michel Foucault's 'Panopticon' revisited. *Theoretical Criminology* 1: 215–34.

Maydaily.com 2011, August 11: TV anchor denies being detained. We believe her. Retrieved from http://www.maydaily.com/2011/08/11/tv-anchor-denies-being-detained-we-believe-her/

McCaughey, M. and Ayers, M. D. (eds.) 2003: *Cyberactivism: Online Activism in Theory and Practice*. London and New York: Routledge.

McClelland, J. S. 1989: *The Crowd and the Mob: From Plato to Canetti*. London: Unwin Hyman.

Meikle, G. 2002: *Future Active: Media Activism and the Internet*. New York: Routledge.

Meng, B. 2009: Regulation *e'gao*: Futile efforts of recentralization? In X. Zhang and Y. Zheng (eds.), *China's Information and Communications Technology Revolution: Social Changes and State Responses*. London and New York: Routledge, 52–67.

Minjian chunwan民间春晚 (Folk Spring Festival Gala) (2010, February 17). Retrieved from http://www.tudou.com/programs/view/WbrfW_bSyNg/

Minzner, C. 2006: Xinfang: An alternative to formal Chinese legal institutions. *Stanford Journal of International Law* 42, 103–79.

Mirzoeff, N. 2011, October 26: Occupy theory. Retrieved from http://critinq.wordpress.com/2011/10/26/occupy-theory/

Moore, M. 2009, December 30: Chinese Internet vigilantes bring down another official. Retrieved from http://www.telegraph.co.uk/news/worldnews/asia/china/4026624/Chinese-internet-vigilantes-bring-down-another-official.html

Morris, M. 2001, Contradictions of postmodern consumerism and resistance. *Studies in Political Economy* 64, 7–29.

Morris, D. 2004: Globalization and media democracy: The case of Indymedia. In D. Schuler and P. Day (eds.), *Shaping the Network Society: The New Role of Civil Society in Cyberspace*. Cambridge, MA: MIT Press, 325–52.

Naughton, J. 2001: Contested space: The Internet and global civil society. In H. Anheier., M. Glasius and M. Kaldor (eds.), *Global Civil Society 2001*, Oxford: Oxford University Press, 147–68.

News.163.com 2008, December 16: Wangyou bao Nanjing yi fangchan juzhang chou tianjiayan 1500/tiao 网友曝南京一房产局长抽天价烟1500/条 (Netizen reveals a director of the Property Bureau in Nanjing smokes luxury cigarettes at 1500 *yuan* per carton). Retrieved from http://news.163.com/08/1216/05/4T8S3EQ600011229.html

News.cntv.cn 2010, February 14: Zhongguo wangluo dianshitai chunwan zhibo huode haineiwai haoping 中国网络电视台春晚直播获得海内外好评 (China Network Television's live broadcast of Spring Festival Gala received favorable comments from home and abroad). Retrived from http://news.cntv.cn/china/20100214/100291.shtml

News.qq.com 2008, January 17: Wu xuezhe lianming huyu dizhi chunwan, cheng weiminsu xiedu chuantong 五学者联名呼吁抵制春晚, 称伪民俗亵渎传统 Five scholars collectively resist CCTV's Spring Festival Gala, claiming faked folk customs profane tradition). Retrieved from http://news.qq.com/a/20080117/002407.htm

News.qq.com 2009a, February 18: Shiping manhua: kanshousuo "duomaomao" dasai 时评漫画：看守所"躲猫猫"大赛 (Commentary cartoon: "Hide-and-seek" competition in detention centre). Retrieved from http://news.qq.com/a/20090218/000432.htm

News.qq.com 2009b, February 20: Yunnan yaoqing wangmin diaocha duomaomao shijian, 15 ren mingdan gongbu 云南邀请网名调查躲猫猫事件，15人名单公布 (Yunnan invites netizens to investigate the duomaomao incident: a name list of 15 people is released). Retrieved from http://news.qq.com/a/20090220/000071.htm

News.qq.com 2015, January 22: Gongxinbu: 2014 nian zhongguo shouji yonghu yida 12.86 yihu工信部：2014 年中国手机用户已达12.86亿户 (Ministry of Industry and Information Technology: China's mobile phone users reached 1.286 billion in 2014). Retrieved from http://news.qq.com/a/20150122/010637.htm

News.sina.com.cn 2014, July 24: 2014 nian shangbannian xinlang zhengwu Weibo baogao jinri fabu 2014 年上半年新浪政务微博报告今日发布 (Sina Official Weibo Report for first half of 2014 released today). Retrieved from http://news.sina.com.cn/c/2014-07-24/111830571953.shtml

News.souhu.com 2009, February 22: Wangmin diaocha duomaomao beizhi zuoxiu, diaochatuan chengyuan bei renrou sousuo 网民调查躲猫猫被指作秀, 调查团成员被人肉搜索 Netizens' investigation into "hide-

and-seek" incident suspected of being a show: Members of the investigation team were hunted online). Retrieved from http://news.sohu.com/20090222/n262384023.shtml

News.xinhuanet.com 2015, January 29: 2014 zhengwu zhishu baogao fabu, zhengwu Weibo chengwei "xinchangtai" 《2014政务指数报告》发布，政务微博成为"新常态" (2014 Government Administration Index Report released: Official Weibo has become the new norm). Retrieved from http://news.xinhuanet.com/zgjx/2015-01/29/c_133955732.htm

Nip, J. 2009: Citizen journalism in China's Wenchuan earthquake. In S, Allan and E. Thorsen (eds.), *Citizen Journalism: Global Perspectives*, Volume 1, New York: Peter Lang, 95–106.

Noesselt, N. 2014: Microblogs and the adaptation of the Chinese party-state's governance strategy. *Governance* 27 (3), 449–68.

Outing, S. 2005, January 6: Taking Tsunami coverage into their own hands. Retrieved from http://www.poynter.org/uncategorized/29330/taking-tsunami-coverage-into-their-own-hands/

Pan, M., Ling, H. and Yu, C. 2007: Guoneiwai BBS luntan fazhan ji guanli bijiao yanjiu 国内外BBS论坛发展及管理比较研究 (A comparative study of the development and management of BBS forums in China and abroad). *Sixiang lilun jiaoyu daokan* 思想理论教育导刊 (*Journal of Ideological & Theoretical Education*), 7, 68–72.

Pan, P. 2006, January 24: Leading publication shut down in China. Retrieved from http://www.washingtonpost.com/wp-dyn/content/article/2006/01/24/AR2006012401003.html

Pan, X. 2010: Hunt by the crowd: An exploratory qualitative analysis on cyber surveillance in China. *Global Media Journal* 9 (16), 1–19.

Pan, Z. 2000: Improvising reform activities: The changing reality of journalistic practice in China. In C. Lee (ed.), *Power, money and media: Communication patterns and bureaucratic control in cultural China*, Evanston, Illinois: Northwestern University Press, 68–111.

Pan, Z. 2010a: Enacting the family-nation on a global stage: An analysis of CCTV's Spring Festival Gala. In M. Curtin and H. Shah (eds.), *Reorienting Global Communication: Indian and Chinese Media Beyond Borders*, Urbana: University of Illinois Press, 240–59.

Pan, Z. 2010b: Articulation and re-articulation: Agendas for understanding media and communication in China. *International Journal of Communication* 4, 517–30.

People.com.cn 2009, February 20: Jinri huati: guanfang zheng wangmin diaocha "duomaomao" zhenxiang 今日话题：官方征网民调查"躲猫猫"真相 (Today's topic: government recruits netizens to investigate the truth of the "Hide-and-Seek" incident). Retrieved from http://opinion.people.com.cn/GB/8844421.html

People's Daily 2010, February 8: Shanzhai CCTV Spring Festival Gala to go online. Retrieved from http://english.peopledaily.com.cn/90001/90782/6889924.html

Peretti, J. 2001, April 9: My Nike media adventure. Retrieved from http://www.thenation.com/article/my-nike-media-adventure

Perry, E. J. 2002: Moving the masses: Emotion work in the Chinese revolution. *Mobilization: An International Journal* 7 (2), 111–28.

Pickard, V. W. 2006a: Assessing the radical democracy of Indymedia: Discursive, technical, and institutional constructions. *Critical Studies in Media Communication* 23 (1), 19–38.

Pickard, V. W. 2006b: United yet autonomous: Indymedia and the struggle to sustain a radical democracy network. *Media, Culture & Society* 28 (3), 315–36.

Pierson, D. 2011, February 26: Online call for protests in China prompts crackdown. Retrieved from http://articles.latimes.com/2011/feb/26/world/la-fgw-china-crackdown-20110227

Pilger, J. 2005: *Tell Me No Lies: Investigative Journalism that Changed the World*. New York: Thunder's Mouth Press.

Price, M. E. and Dayan, D. (eds.) 2008: *Owning the Olympics: Narratives of the New China*. Ann Arbor, MI: Michigan University Press.

Qian, G. 1986: *Tangshan dadizhen* 唐山大地震 (*Tangshan Earthquake*). Beijing: Jiefangjun wenyi chubanshe 解放军文艺出版社.

Qian Gang's Weibo (2011, July 25–26). Retrieved from http://weibo.com/1995801167

Qian, Y. and Xu. C. 1993: Why China's economic reform differ: The m-form hierarchy and entry/expansion of the non-state sector. *Economics of Transition* 1 (2), 135–70.

Qiu, J. L. 2006: The changing web of Chinese nationalism. *Global Media and Communication* 2 (1), 125–28.

Qiu, J. L. 2009: *Working-class Network Society: Communication Technology and the Information Have-less in Urban China*. Cambridge, MA: The MIT Press.

Qiu, J. L. and Chan, J. 2011: Maixiang xinmeiti shijian yanjiu 迈向新媒体事件研究 (Studies on new media events). In J. L. Qiu and J. Chan (eds.), *Xinmeiti shijian yanjiu* 新媒体事件研究 (*Studies on New Media Events*), Beijing: Renmin daxue chubanshe 人民大学出版社, 1–16.

Qiu, J. L. and Chan, J. 2011: *Xinmeiti shijian yanjiu* 新媒体事件研究 (*Studies on New Media Events*). Beijing: Renmin daxue chubanshe 人民大学出版社.

Qu, Y., Wei, P. and Wang, X. 2009: Online community response to major disaster: A study of Tianya forum in the 2008 Sichuan earthquake. Paper presented at the 42nd Hawaii International Conference on System Sciences, January 5–8, Hawaii. Retrieved from http://ieeexplore.ieee.org/xpl/mostRecentIssue.jsp?punumber=4755313

Radio Free Asia 2009, October 21: Ai Xiaoming boke beifeng. Xianggang tuanti jiaying women de wawa 艾晓明博客被封，香港团体加映《我们的娃娃》(Ai Xiaoming's blog shut down; Hong Kong shows *Our Children* in cinema). Retrieved from http://www.rfa.org/mandarin/yataibaodao/aixiaoming-10212009134620.html

Ramzy, A. 2011a, February 21: State stamps out small Jasmine protests in China. Retrieved from http://www.time.com/time/world/article/0,8599,2052860,00.html

Ramzy, A. 2011b, July 28: After train crash, Chinese P.M. Wen Jiabao joins the government's stumbling P.R. effort. Retrieved from http://globalspin.blogs.time.com/2011/07/28/after-train-crash-chinese-p-m-wen-jiabao-joins-the-governments-stumbling-p-r-effort/

Ran, Y. 2007, December 25: Minkan zai zhongguo de yiyi 民刊在中国的意义 (The signficance of non-governmental periodicals in China). Retrieved from http://2newcenturynet.blogspot.com/2007/12/blog-post_25.html

Rednet.cn 2009, February 19: Guanyu canyu diaocha duomaomao yulun shijian zhenxiang de gonggao 关于参与调查躲猫猫舆论事件真相的公告 (Announcement on taking part in the investigation of duomaomao incident). Retrieved from http://gov.rednet.cn/c/2009/02/19/1710932.htm

Reese, S. D. and Dai, J. 2009: Citizen journalism in the global new arena: China's new media critics. In S. Allan and E. Thorsen (eds.), *Citizen Journalism: Global Perspectives*, Volume 1, New York: Peter Lang, 221–32.

Regan, T. 2003 Weblogs threaten and inform traditional journalism. Retrieved from http://www.nieman.harvard.edu/reports/article/101041/Weblogs-Threaten-and-Inform-Traditional-Journalism.aspx

Reuters.com 2011, August 23: China rail firm boss, blamed for crash, dies of heart attack. Retrieved from http://www.reuters.com/article/2011/08/23/us-china-railways-idUSTRE77M2H820110823

Ricketson, M. 2001, August 20: The importance of investigative journalism—of journalism. Retrieved from http://www.abc.net.au/4corners/4c40/essays/ricketson.htm

Roberts, I. 2010: China's Internet celebrity: Furong jiejie. In L, Edwards and E. Jeffreys (eds.), *Celebrity in China*. Hong Kong: Hong Kong University Press, 217–36.

Rodriguez, C. 2011: *Citizens' Media Against Armed Conflict: Disrupting Violence in Colombia*. Minneapolis, MN: University of Minnesota Press.

Rosen, J. 2006, June 27: The people formerly known as the audience. Retrieved from http://archive.pressthink.org/2006/06/27/ppl_frmr.html

Rothenbuhler, E. W. 1998: The living room celebration of the Olympic Games. *Journal of Communication* 38 (3), 61–81.

Rudé, G. 1964: *The Crowd in History: A Study of Popular Disturbances in France and England, 1730–1848*. New York: Wiley.

Ryfe, D. and Mensing, D. 2008: Participatory journalism and the transformation of news. Paper presented at the annual meeting on the Association for Education in Journalism and Mass Communication, August 6, Chicago. Retrieved from http://citation.allacademic.com/meta/p_mla_apa_research_citation/2/7/1/5/8/pages271585/p271585-1.php

Scannell, P. 1996: *Radio, Television and Modern Life*. Cambridge: Blackwell.

Scannell, P. 1999: The death of Diana and the meaning of media events. *Review of Media, Information and Society* 4, 27–50.

Scannell, P. 2002: Big brother as television event. *Television and New Media* 3 (3), 271–82.

Schurmann, F. 1968: *Ideology and Organization in Communication China*. Berkeley, CA: University of California Press.

Schwartz, E. 1996: *NetActivism: How Citizens use the Internet*. Sebastopol, CA: O'Reilly Media.

Sellnow, T. L., Seeger, M. W. and Ulmer, R. R. 2002: Chaos theory, informational needs and natural disasters. *Journal of Applied Communication Research* 30 (4), 269–92.

Shanzhai chunwan 山寨春晚 (Spring Festival Gala) (2009, January 28). Retrieved from http://www.youtube.com/watch?v=rCWglyul7fA&feature=PlayList&p=147B058ACA9DE719&playnext_from=PL&index=0&playnext=1

Shanghai jimo xiaosheng's Weibo (2011, July 27). Retrieved from http://www.weibo.com/1617464310

Shen, S. and Breslin, S. (eds.) 2010: *Online Chinese Nationalism and China's Bilateral Relations*. Lanham: Lexington.

Sheng, H. 1990: Big character posters in China: A historical survey. *Journal of Chinese Law* 4(2), 234–56.

Shibutani, T. 1966: *Improvised News: A Sociological Study of Rumor*. Indianapolis, IN: Bobbs-Merrill.

Shields, M. 2011, March 28: Twitter co-founder Jack Dorsey rejoins company. Retrieved from http://www.bbc.co.uk/news/business-12889048

Shirky, C. 2008: *Here Comes Everybody: The Power of Organizing Without Organizations*. London: Allen Lane/Penguin.

Shisan hu's blog (2009, February 24). Retrieved from http://blog.sina.com.cn/s/blog_49b457eb0100bp98.html

Shrivastava, P. 1987: *Bhopal: Anatomy of a Crisis*. Cambridge, MA: Ballinger Publishing Company.

Silverstone, R. 2005: The sociology of mediation and communication. In C. Calhoun, C. Rojek and B.S. Turner (eds.), *The Sage Handbook of Sociology*, Thousand Oaks, CA: Sage, 188–207.

Sohu.com 2003, April 21: Guowuyuan xinwenban guanyu feidian jizhe zhaodaihui 国务院新闻办关于非典记者招待会 (State Council Information Office's press conference about SARS). Retrieved from http://health.sohu.com/00/25/harticle17342500.shtml

Sohu.com 2008, December 5: Shanzhai chunwan huo qiye rongzi baiwan wangyou baoming jiemu 500 duoge 山寨春晚获企业融资百万 网友报名节目500 (Shanzhai Spring Festival Gala receives 1 million yuan in financing from business; Netizens recommend more than 500 programs). Retrieved from http://yule.sohu.com/20081205/n261029897.shtml

Southern Metropolitan Daily 2009: Wangluo daiyou xincichu zuijin liuxing "duomaomao" 网络代有新词出最近流行"躲猫猫" ("Duomaomao" became the latest online buzzword). February 16: AA24.

Spencer, R. 2008, June 7: China orders journalists to end negative quake coverage. Retrieved from http://www.telegraph.co.uk/news/worldnews/asia/china/2091084/China-earthquake-journalists-orderered-to-end-negative-quake-coverage.html

Statista 2015: Number of monthly active WeChat users from 2nd quarter 2010 to 1st quarter 2015 (in millions). Retrieved from http://www.statista.com/statistics/255778/number-of-active-wechat-messenger-accounts/

Strömbäck, J. 2008: Four phases of mediatization: An analysis of the mediatization of politics. *The International Journal of Press/Politics* 13 (3), 331–45.

Strömbäck, J. and Esser, F. 2009: Shaping politics: Mediatization and media interventionism. In K. Lundby (ed.), *Mediatization: Concept, Changes and Consequences*. New York: Peter Lang, 205–224.

Sturges, D. L. 1994: Communicating through crisis: A strategy for organizational survival. *Management Communication Quarterly* 7 (3), 297–316.

Su, S. 1994: Chinese communist ideology and media control. In C. C. Lee (ed.), *China's Media, Media's China*. Boulder, CO: Westview Press, 75–88.

Sullivan, J. 2013: China's Weibo: Is faster different, *New Media & Society* 16 (1), 24–37.

Sun, F. 2001: Cong "renbenwei" dao "shibenwei": woguo zainan baodao bianhua fenxi "人本位"到"事本位"：我国灾难报道变化分析 (From "people-oriented" to "fact-oriented": Analysis of changes in disaster news in China). *Xiandai chuanbo* 现代传播 (*Modern Communication*) 109, 33–37.

Sun, W. 2007: Dancing with chains: Significant moments on China Central Television. *International Journal of Cultural Studies* 10 (2), 187–204.

Sun, W. 2010: Mission impossible? Soft power, communication capacity, and the globalization of Chinese media. *International Journal of Communication* 4, 54–72.

Sun, W. 2014: Media events: Past, present and future. *Sociology Compass* 8 (5), 457–67.

Sydney Morning Herald 2003, December 1: Net sex writer stirs emotions. Retrieved from http://www.smh.com.au/articles/2003/12/01/1070127357698.html

Tech.qq.com 2009, January 7: Guonei shouji xiaoliang disan jidu xiajiang 2.7% 国内手机销量第三季度下降2.7% (The sales volume of mobile phones in the domestic market decreases 2.7% in the third season). Retrieved from http://tech.qq.com/a/20090107/000011.htm

Teng, B. 2012: Rights defence (*weiquan*), microblogs (w*eibo*), and popular surveillance (*weiguan*). *China Perspectives* (3), 29–39.

Thompson, J. B. 1997: Scandals and social theory. In J. Lull and S. Hinerman (eds.), *Media Scandals: Morality and Desire in the Popular Culture Marketplace,* New York: Columbia University Press, 34–64.

Thurman, N. and Rodgers, J. 2014: Citizen journalism in real time: Live blogging and crisis events. In E. Thorsen and S. Allan (eds.), *Citizen Journalism: Global Perspectives*, Volume 2, New York: Peter Lang, 81–95.

Toffler, A. 1980: *The Third Wave*. New York: Morrow.

Tong, J. 2011: *Investigative Journalism in China: Journalism, Power and Society*. New York and London: Continuum.

Tong, J. and Sparks, C. 2009: Investigative journalism in China today. *Journalism Studies* 10 (3), 337–52.

Tran, T. 2008, May 15: Media given green light to report on disaster. Retrieved from http://www.thestar.com/news/world/2008/05/15/media_given_green_light_to_report_on_disaster.html

Tse, E., Ma, K. and Huang, Y. 2009: Shanzhai: A Chinese phenomenon. Retrieved from http://www.booz.com/media/file/Shan_Zhai_A_Chinese_Phenomenon_en.pdf

Tsui, L. 2003: The panopticon as the antithesis of a space of freedom: Control and regulation of the Internet in China. *China Information: A Journal on Contemporary China Studies* 17 (2), 65–82.

Van Laer, J. and Aelst, P. 2010: Internet and social movement action repertoires: Opportunities and limitations. *Information, Communication & Society* 13 (8), 1146–71.

Venturedata.org 2013, January 25: CCTV Spring Festival Evening "zero advertising" three years less earned 1.75 billion? Retrieved from http://www.venturedata.org/?i480777_CCTV-Spring-Festival-Evening-zero-advertising-three-years-less-earned-1.75-billion

Viviani, M. 2010: Chinese independent documentary films: What role in contemporary China? Paper presented at the 18th Biennial Conference of the Asian Studies Association of Australia, July 5–8, Adelaide.

Retrieved from http://asaa.asn.au/ASAA2010/reviewed_papers/Viviani-Margherita.pdf

Voci, P. 2010: *China on Video: Smaller-screen Realities*. London and New York: Routledge.

Volkmer, I. 2008: Conflict-related media events and cultures of proximity. *Media, War & Conflict* 1 (1), 90–98.

Waisbord, S. 2004: Scandals, media and citizenship in contemporary Argentina. *The American Behavioral Scientist* 47 (8), 1072–98.

Wall, M. 2009: The taming of the warblogs: Citizen journalism and the war in Iraq. In S. Allan. and E. Thorsen (eds.), *Citizen Journalism: Global Perspective*, Volume 1, New York: Peter Lang, 33–42.

Waltz, M. 2005: *Alternative and Activist Media*. Edinburgh: Edinburgh University Press.

Wang. D. 2008: Laolao bawo xinwen xuanchuan gongzuo de zhudongquan 牢牢把握新闻宣传工作的主动权 (Firmly grasp the initiative in news and propaganda work). Retrieved from http://xwjz.eastday.com/eastday/xwjz/node271090/node271091/u1a3694705.html

Wang, G. 2010, February 9: China sentences activist who investigated children's deaths in 2008 quake to 5 years' jail. Retrieved from http://www.startribune.com/templates/Print_This_Story?sid=83850617

Wang, Q. and Fan, Y. 2014: Weixin boxing, Weibo yiran buke bei tidai 微信勃兴，微博依然不可被替代 (Weibo remains irreplaceable with the rise of Weixin). *Xinwen jizhe* 新闻记者 (*Shanghai Journalism Review*) 4, 59–62.

Wang, X. 2009, May 18: 'Shanzhai' culture now in crosshairs. Retrieved from http://www.chinadaily.com.cn/bw/2009-05/18/content7785393.htm

Wang, X. B. 2009, August 19: Zai gonggong tufa shijian baodao zhong tixian dianshi xinwen jigou de baqi 在公共突发事件报道中体现电视新闻机构的霸气 (Television media's leadership in reporting crises). Retrieved from http://media.people.com.cn/GB/22114/70684/166538/9891041.html

Warren, M. E. 2009: Governance-driven democratization. *Critical Policy Studies* 3 (1), 3–13.

Watts, J. 2006, December 6: Outrage at Chinese prostitutes' shame parade. Retrieved from http://www.guardian.co.uk/world/2006/dec/06/china.jonathanwatts

Wen, Y. 2009: Women de yizhi shi leguande: Zhongguo linglei chuanbo de shengji jiuzai jiafengzhong我们的意志是乐观的：中国另类传播的生机就在夹缝中 (We are optimistic: China's alternative media live in fissures). *Xinwenxue yanjiu* 新闻学研究 (*Journalism Studies*) 99, 251–64.

Wu, A. 2009, April 28: Wenchuan zhihou, shui sizai dierci 汶川之后，谁死在第二次？(After the Sichuan earthquake, who will die the second time?) Retrieved from http://www.nfcmag.com/article/1485.html

Wu, X. 2006, February 6: Shanzhai wenhua tuxian caogen jingshen 山寨文化凸显草根精神 (*Shanzhai* culture signifies grassroots spirit). Retrieved from http://wenku.baidu.com/view/171bb671a417866fb84a8e3e.html

Wu, X. 2007: *Chinese Cyber Nationalism: Evolution, Characteristics, and Implications*. Lanham: Lexington.

Xi, W. 2009: The recurring shanzhai: A phenomenon. *China Today* 58, 42–46.

Xiaodan tongxue qiu shunli's Weibo (2011, July 28). Retrieved from http://www.weibo.com/snowbeardan

Xie, Y., Cao, Z. and Wang, T. 2009: *Tufa shijian baodao* 突发事件报道 (*Reporting on Emergencies*). Shanghai: Shanghai Jiaotong daxue chubanshe 上海交通大学出版社.

Xin, X. 2010: The impact of "citizen journalism" on Chinese media and society. *Journalism Practice* 4 (3), 333–44.

Xinhuanet.com 2003a, July 3: Yixie guojia luxu jiechu duiwo wangfang tuanzu he renyuan de xianzhi cuoshi 一些国家陆续解除对我往访团组和人员的限制措施 (Selected countries lift restrictions on Chinese delegations and visitors). Retrieved from http://news.xinhuanet.com/zhengfu/2003-07/03/content_952228.htm

Xinhuanet.com 2003b, September 27: Feidian jiakuai wanshan xinwen fabu zhidu 非典加快完善新闻发布制度 (The SARS epidemic has improved China's news release system). Retrieved from http://news.xinhuanet.com/newscenter/ 2003-09/27/content_1102144.htm

Xinhuanet.com 2005, February 2: Chunwan fei hanyu fanyi gongzuo wancheng, zhiboshi sizhong yuyan qishangzhen春晚非汉语翻译工作完成，直播时四种语言齐上阵 (Spring Festival Gala translated into languages other than Chinese; will be broadcast live in four foreign languages. Retrieved from http://news.xinhuanet.com/ent/2005-02/02/content_2537651.htm

Xinhuanet.com 2006, January 8: Guojia tufa gonggong shijian zongti yingji yu'an 国家突发公共事件总体应急预案 (National emergency response plan). Retrieved from http://news.xinhuanet.com/politics/2006-01/08/content_4024011.htm

Xinhuanet.com 2009a, January 26: Chuxiye yangshi chunwan shoushilü chaoguo 95% 除夕夜央视春晚收视率超过95% (CCTV Spring Festival Gala attracts over 95% of viewers). Retrieved from http://news.xinhuanet.com/newscenter/2009-01/26/content_10723086.htm

Xinhuanet.com 2009b, May 11: Zhongguo 1300 duo wan ming zhiyuanzhe canyu wenchuan dizhen kangzhen jiuzai 中国1300多万名志愿者参与汶川地震抗震救灾 (More than 13 million volunteers participate in rescue and relief work after Wenchuan earthquake). Retrieved from http://news.xinhuanet.com/newscenter/2009-05/11/content_11351632.htm

Xinhuanet.com 2010, January 5: Bei "zhengtong" zhao'an, shangyehua de caogen chunwan haineng zou duoyuan? 被"正统"招安，商业化的草根春晚还能走多远？ (Accepting the authorities' amnesty, how far can the commercialized Spring Festival galas go?) Retrieved from http://news.xinhuanet.com/focus/2010-01/05/content_12757057.htm

Xu, X. 2006, July 28: "Chidao de xinwen": Tangshan dizhen siwang renshu baodao beihou "迟到的新闻"：唐山地震死亡人数报道背后 (Belated news: Behind the death toll of the Tangshan earthquake). Retrieved from http://news.xinhuanet.com/newmedia/2006-07/28/content_4888384.htm

Xue, L. and Liu, B. 2013, June 17: Pandian "feidian" shinian: gonggong zhili tixi biange 盘点"非典"十年：公共治理体系变革 (Taking stock on the 10th anniversary of SARS: Transformation of public governance system). Retrieved from http://theory.people.com.cn/n/ 2013/0617/c49154-21866221.html

Yang, G. 2003: The co-evolution of the Internet and civil society in China. *Asian Survey* 43, 405–22.

Yang, G. 2008: Contention in Chinese cyberspace. In K. O'Brien (ed.), *Popular Protest in China*, Cambridge, MA and London: Harvard University Press, 126–43.

Yang, G. 2009a: *The Power of the Internet in China: Citizen Activism Online*. New York: Columbia University Press.

Yang, G. 2009b: Online activism. *Journal of Democracy* 20 (3), 33–36.

Yang, G. 2009c: Historical imagination in the study of Chinese digital civil society. In X. Zhang and Y. Zheng (eds.), *China's Information and Communications Technology Revolution: Social Changes and State Responses*. London and New York: Routledge, 17–33.

Yang, G. 2014: The return of ideology and the future of Chinese Internet policy. *Critical Studies in Media Communication* 31 (2), 109–13.

Ye, S. 2011, February 14: Sina Weibo adds voicemail Weibos and direct video uploads. Retrieved from http://techrice.com/2011/02/14/sina-weibo-adds-voicemail-weibos-and-direct-video-uploads/

Yin, L. and Wang, H. 2010: People's myth: A discourse analysis of Wenchuan earthquake in *China Daily*. *Discourse and Communication* 4 (4), 383–98.

Yu, F. T. C. 1964: *Mass Persuasion in Communist China*. New York: Praeger.

Yu, H. 2006: From active audience to media citizenship: The case of post-Mao China. *Social Semiotics* 16 (2), 303–26.

Yu, H. 2007a: Blogging everyday life in Chinese Internet culture. *Asian Studies Review* 31 (4), 423–33.

Yu, H. 2007b: Talking, linking, clicking: The politics of AIDS and SARS in urban China. *Positions: East Asia Cultures* Critique 15 (1), 35–63.

Yu, H. 2009: *Media and Cultural Transformation in China*. London and New York: Routledge.

Yu, H. 2011: *Dwelling Narrowness*: Chinese media and their disingenuous neoliberal logic. *Continuum: Media & Cultural Studies* 25 (1), 33–46.

Yuan Xiaoyuan's Weibo (2011, July 23). Retrieved from http://weibo.com/profile.php?uid=1144332832&page=8

Zhang, J. 2010, January 24: Ai Xiaoming de chuanzhen jilupian zhenhan 艾晓明的川震纪录片震撼 (The shock of Ai Xiaoming's documentaries on the Sichuan earthquake). Retrieved from http://www.yzzk.com/cfm/Content_Archive.cfm?Channel=ag&Path=3236035881/04ag2a.cfm

Zhang, L. 2006: Behind the 'Great Firewall': Decoding China's Internet media policies from the inside. *Convergence: The International Journal of Research into New Media Technologies* 12 (3), 271–91.

Zhang, X. 2011: *The Transformation of Political Communication in China: From Propaganda to Hegemony*. Singapore and London: World Scientific.

Zhao, B. 1998: Popular family television and Party ideology: The Spring Festival Eve happy gathering. *Media, Culture and Society* 20 (1), 43–58.

Zhao, Y. 1997: Toward a propagandist/commercial model of journalism in China? The case of the Beijing Youth News. *Gazette: The International Journal of Communication Studies* 58 (3), 143–57.

Zhao, Y. 1998: *Media, Market, and Democracy in China: Between the Party Line and the Bottom Line*. Urbana and Chicago, IL: University of Illinois Press.

Zhao, Y. 2000: Watchdog on Party leashes? Contexts and implications of investigative journalism in post-Deng China. *Journalism Studies* 1 (2), 577–97.

Zhao, Y. 2008a: *Communication in China: Political Economy, Power and Conflict*. Lanham, MD: Rowman & Littlefield.

Zhao, Y. 2008b: Neoliberal strategies, socialist legacies: Communication and state transformation in China. In P. Chakravartty and Y. Zhao (eds.), *Global Communication: Towards a Transcultural Political Economy*, Lanham, MD: Rowman & Littlefield, 23–50.

Zheng, Y. 2008: *Technological Empowerment: The Internet, the State and Society in China*. Stanford, CA: Stanford University Press.

Zheng, Y. and Wu. G. 2005: Information technology, public space, and collective action in China. *Comparative Political Studies* 38 (5), 507–36.

Zhou Shuguang's blog (2008, May 15–21). Retrieved from https://www.zuola.com/weblog/sitemap?pg=2

Zhu, S. and Shi, Y. 2010: *Shanzhai* manufacturing—an alternative innovation phenomenon in China: Its value chain and implications for

Chinese science and technology policies. *Journal of Science and Technology Policy in China* 1 (1), 29–49.

Zhu, W. 2005, July 18. Meiti dui feidian shijian baodao qingkuang de diaocha baogao 媒体对非典事件报道情况的调查报告 (A report on media's coverage of the SARS epidemic). Retrieved from http://www.zijin.net/blog/user1/247/archives/2005/2505.shtml.

Index

Page numbers in **bold** refer to tables, those in *italic* to illustrations. *Italic* text in entries is used for names of publications and other resources, artists' works and for Chinese terms.

Printed and bound by CPI Group (UK) Ltd, Croydon, CR0 4YY

13/09/2026

14961102-0002